NONFUEL MINERALS

NONFUEL MINERALS
Foreign Dependence and National Security

A Twentieth Century Fund Book

Raymond F. Mikesell

The University of Michigan Press
Ann Arbor

Published in the United States of America by
The University of Michigan Press and simultaneously
in Markham, Canada, by Fitzhenry & Whiteside, Ltd.
Manufactured in the United States of America
1990 1989 1988 1987 4 3 2 1

Library of Congress Cataloging-in-Publication Data

Mikesell, Raymond Frech.
Nonfuel minerals.

"A Twentieth Century Fund book."
Bibliography: p.
Includes index.
1. Nonfuel minerals industry—United States.
2. Nonfuel minerals industry—Government policy—
United States. 3. Strategic materials—Government
policy—United States. I. Title.
HD9506.U62M42 1987 333.8′5′0973 86-27247
ISBN 0-472-10083-1 (alk. paper)

Yet another book for Irene, who has inspired them all

The Twentieth Century Fund is an independent research foundation which undertakes policy studies of economic, political, and social institutions and issues. The Fund was founded in 1919 and endowed by Edward A. Filene.

Foreword

The oil crisis of the 1970s led the federal government to enact a series of emergency measures, some of which proved wasteful and expensive mistakes. That experience contains useful lessons for assuring reasonably priced supplies of nonfuel mineral resources. But we have been curiously lethargic about formulating a sensible long-term policy to deal with nonfuel minerals. Part of the reason is that whenever we have been faced in the past with shortages of critical nonfuel minerals, we have always managed to cope. Our ad hoc approach to stockpiling and our vaunted ingenuity in finding substitutes has seen us through. But in the past we were able to count on a steady flow of supplies from South Africa. Now, however, the growing hostility toward the white minority government of South Africa and its abhorrent racist policy of apartheid has raised the specter of critical shortages of some materials, such as cobalt, chromium, and the platinum-group metals, vital for our defense industry and for parts of our increasingly high-tech economy.

Even if the United States embargoes all imports from South Africa, we will probably find ways of muddling through. But the costs could be high, and the prospect of production stoppages cannot be ruled out. Surely it would be wiser to consider policies that guard against worst-case scenarios. By the same token, we ought to be considering how to protect our economy against the disruption of needed imports from those Third World suppliers that have been suffering from declining prices in recent years.

The need for an examination of these policy issues was brought to the attention of the Twentieth Century Fund by Raymond Mikesell, a respected senior economist who has devoted much of his long professional career to the analysis of domestic and international nonfuel resources. He proposed a comprehensive examination of the question of how best to bring about a reliable and reasonably priced flow of nonfuel minerals. Since the Fund has had a traditional interest in this policy area, we welcomed the opportunity to sponsor this needed research.

In his report, Dr. Mikesell assesses current national policy regarding each of the major nonfuel minerals, examining existing stockpiling programs and appraising the current sources for each mineral and the threat of a disruption in their flow to the United States. He then evaluates the costs and availability of potential substitutes. In addition, Dr. Mikesell focuses on critical environmental issues, dealing in particular with the role of domestic mining interests and the government officials, state and federal, who are engaged in monitoring and regulating mining. A sophisticated analyst, he evaluates the trade-offs between protecting the environment and maintaining or increasing production, seeking out the most balanced and fairest course.

Despite the occasional flurry of interest in this subject, it is for the most part neglected by policymakers and the public except when we are faced, as we seem to be at the moment, by the threat of shortages. We are grateful to Dr. Mikesell for his painstaking analysis, and especially for pointing out the costs and dangers in a policy of drift. He argues for a more coherent policy that takes into account environmental and economic concerns and seeks to avoid undue risks. We think that his views merit both consideration and debate.

M. J. Rossant, Director
The Twentieth Century Fund

Preface

The familiar prophecy that world economic growth will soon be limited by the exhaustion of mineral resources is seldom heard today. What is heard is that overcapacity and low prices will destroy America's mineral industries and leave the country vulnerable to import disruptions resulting from a major war or from civil disturbance in South Africa or elsewhere. When prices of minerals soar again, as I believe they will some day, the doomsday prophecies of a decade and a half ago will be heard again. This book attempts to allay both past and present fears. It is based on the premise that free and competitive world markets and technological progress can assure both abundant minerals and national security without government interference with production and trade.

Nevertheless, regulations are needed to protect the environment, since our air, water, and land constitute an indispensable resource that must not be allowed to deteriorate. We also need to stockpile minerals—happily few—not produced on this continent as insurance against world calamities. But our economic welfare and national security will not be served by government subsidies and trade restrictions on nonfuel minerals or by national or international schemes to fix their prices.

My interest in the subject of this book dates from my service on the staff of the President's Materials Policy (Paley) Commission in 1951. When the commission was appointed by President Harry Truman, the U.S. government feared that world supplies of minerals might soon be exhausted and that a national program was required to assure their future availability. Before the commission's report was prepared, the demand-supply balance for most minerals had improved so markedly that the commission and its staff not only doubted the need for a comprehensive government program, but questioned whether any report on minerals was justified. Fortunately, the report did not recommend a large government program for minerals. Its analysis and recommendations were sensible but largely ignored.

The ideas expressed in this book were derived from many sources, but the greatest influence has come from three outstanding mineral economists: Hans H. Landsberg of Resources for the Future, Inc.; John E. Tilton of the Colorado School of Mines; and William Vogley of Pennsylvania State University. Its readability is in large measure due to the skillful editing of Bill Bischoff plus the acute sensitivity of my wife, Irene, and my secretary, Letty Fotta, to errors in the manuscript and galleys. Finally, I want to express my wholehearted appreciation to the Twentieth Century Fund for making it possible to undertake the research and preparation of this study.

Contents

1

The National Interest in Nonfuel Minerals: An Unorthodox Assessment

Few Americans—at least since the OPEC boycott of 1973—need to be reminded that the United States, once dominant in world petroleum extraction and marketing, has itself become a net fuel importer, with all the economic and strategic hazards that entails. There has been abundant public debate between oil companies, conservationist groups, and federal and state leadership over the most prudent measures to assure national and economic security without irreparable harm to the environment.

Less well known—and certainly less widely understood—is the fact that prior to World War II this country was not only self-sufficient in most important nonfuel minerals such as steel, copper, and aluminum, but also dominated the world markets for these commodities. Since 1950, however, the international competitive position of the United States in major metals has declined to a point at which the nation is increasingly dependent on foreign sources of supply. The result has been a many-sided but inconclusive dispute over appropriate public policy regarding the production, importation, and adequacy of these vital minerals for the next decade or two.

During the past quarter century, concern over the availability of nonfuel minerals has been focused on the formulation of a comprehensive national minerals policy that would have broad public support. This has been the subject of presidential reports and congressional hearings and legislation; reports of presidential commissions and of governmental agencies, such as the Bureau of Mines, the General Accounting Office, and the Office of Technology Assessment of the U.S. Congress; and publications of nongovernmental organizations, such as the National Materials Advisory Board and the American Mining Congress. These sources have produced a large body of documentation and analysis of considerable technical value but have failed to produce a national minerals policy that goes beyond pious generalizations that compromise a number of conflicting private interests and national objectives. This is not surprising. Each nonfuel mineral, whether iron ore, titanium, or copper, presents particular problems related to basic national goals, such as military security, economic growth, and preservation of social values, that cannot be blanketed by a unified national materials policy. The essential point to emphasize is that pursuit of a national nonfuel minerals policy is an exercise in futility.

Perhaps the best illustration of the chimerical nature of this quest comes from recent federal initiatives beginning with the National Materials and Minerals Policy, Research and Development Act of 1980 (Public Law 96-479). In that law Congress declared it to be United States policy "to promote an adequate and stable supply of materials necessary to maintain national security, economic well-being and industrial production, with appropriate attention to long-term balance between resource production, energy use, a healthy environment, natural resources conservation, and social needs." The president was directed to implement a broad agenda of specific policies and programs, including (within one year of the act's passage) a program, budget proposals, and organizational structure for carrying out its mandates.

President Reagan's response in his statement to Congress of April, 1982, was long on policy compromises with various interest groups and short on specific proposals for federal programs and funding. Major attention went to the availability of public lands for exploration and development of minerals, materials research and development, and the national strategic and critical materials stockpile. The president's message was found wanting in major respects by all its constituencies. For example, it contained something of interest to mining companies and environmentalists alike, but was unsatisfactory as a whole to both. Furthermore, the method of selecting lands for mineral exploration and development gave priority to the existence of mineral values rather than environmental amenities, and it lacked any suggestion of social benefit-cost analysis in the allocation of federal lands among competing uses. In the area of materials research and development, it lacked specific programs for identifying priorities and needs, and it did not put forth any proposals for federal appropriations to fund programs.

The failure to formulate a satisfactory materials policy and a program for its implementation has multiple causes.[1] First, a wide range of commodities is at issue, each with its own economic, military, and social considerations and remedial options. It is therefore hard to see how a common statement and policy applying to all commodities could amount to more than platitudes.[2]

Second, the major national policy goals are not oriented to materials, but have to do with maximizing the growth of the national product: holding down real prices of goods and services, protecting the civilian economy and the national defense capability in periods of emergency, realizing the benefits from international trade, maintaining the health and safety of the community, promoting international peace and economic welfare, and protecting the environment. Materials policies play a role in achieving these national goals, but they are *instruments* to their achievement, not goals themselves.

Third, the principal contributors to the formulation of a national materials policy, including representatives from both government and the private sector, are seldom disinterested groups. The result has not been a national materials policy per se, but a policy statement reflecting the interests of producers and users of minerals. Constituencies devoted to environmental issues, or to the liberalization of international trade, or to the promotion of foreign economic development, have almost no input in the formulation of a national materials policy.

Fourth, at the presidential level, recognition of broad national interests is reflected largely in lip service to a "balanced" materials policy that takes into account conflicting special interests. What is needed, however, is not a "compromise" between contending interests, but rational materials programs designed to promote major national goals.

How, then, can U.S. policymakers best face the dangers that confront the nation—growing dependence on imports of nonfuel minerals from vulnerable sources, such as South Africa and Zaire, and the capture of domestic metal markets by Japan, the European Economic Community (EEC), and the newly industrializing developing countries (NICs) such as Brazil and Korea? Of equal concern to many, can the United States skirt these dangers without wholesale depletion of its remaining nonfuel mineral resources at a destructive cost to environmental values?

No recommendations can be made without systematic investigation of market structure and price fluctuations in the context of a long-term worldwide trend of falling prices for nonfuel minerals and of their technology-derived functional substitutes, and just such an investigation is the focus of the first part of the book. Against this background, the book's second part addresses a number of highly topical public policy issues: U.S. nonfuel mineral production, import dependency, and the case for free trade as opposed to tariff and subsidy protection; competing priorities for environmental protection and wilderness-area preservation vis-à-vis domestic nonfuel mineral production; and safeguards against residual vulnerability to import disruption.

Two points remain to be covered before attacking the larger issues mentioned above: the frequent confusion of national with private economic interests in deciding on nonfuel minerals policy and the method of social benefit-cost accounting to be used here.

Nonfuel mineral issues themselves are sufficiently complex to generate honest differences of opinion among experts, but even more problematic are the public policies adopted or proposed to deal with them. Not only are these public policies frequently based on a misunderstanding or distortion of the issues themselves, but the specter of domestic mineral shortages often serves to conjure up public support for policies heavily weighted toward private economic interests. For example, the national security argument has been enlisted to justify import protection or government subsidy for production from low-grade ore bodies owned by a few mining companies. The national defense argument also serves mining interests that wish to open up designated wilderness areas or national parks for mineral exploration and extraction. Because most of the public regards national defense as more important than any other objective, the defense argument tends to obfuscate rational comparison of social values deriving from environmental protection or from unrestricted international commerce.

The objectives of public policy with regard to nonfuel minerals clearly should not be the guarantee of private profits and continued employment in particular industries. Rather, primary consideration must go to *social* benefits and *social* costs, and to the maximization of *net* social benefits. Social benefits are those accruing to the nation as a whole, and social costs are borne by the nation as a whole, although in both cases the impacts upon various groups (e.g., consumers, workers, and businesses producing or using minerals) will differ. Net social benefits (social benefits minus social costs) are usually reflected in the total national product or income. For example, government actions such as import protection or direct subsidies to domestic mining concerns to make them cost-competitive with foreign imports tend to reduce the real national product. This is true because consumers or taxpayers must bear the resultant higher costs for maintenance of domestic supply. Moreover, domestic capital and labor are employed less efficiently than they could be in other industries, so the national product is thereby reduced.[3]

Some social costs are not readily quantifiable and may not be directly reflected in the national product. For example, the U.S. mineral industry desires both exemption from certain environmental protection regulations related to clean air and water and the right to explore and mine wilderness areas that offer unique environmental amenities. Although mining that reduces air and water quality and destroys wilderness area amenities may increase the national product as it is commonly measured, such activity involves less tangible but nonetheless real social costs, including an indirect decrease in national productivity through impaired health of workers. Thus, there may be trade-offs to consider between increasing the national product as measured and increasing social costs by damaging the environment. Methods have been developed to quantify environmental costs in analyzing these trade-offs to determine net social benefits or losses.

Accordingly, this study employs the principles of social *benefit-cost accounting,* although actual calculations are not made except by way of illustration.[4] Social accounting differs from private accounting in several basic respects. First, social costs constitute net human and material resource costs, including disutilities such as environmental impairment that apply to society as a whole, but social costs do not include income transfers between domestic consumers and producers or between producers and the government.

Second, social benefits and costs include those external to a particular operation or project, while private costs and benefits are confined to those directly involved in a particular operation or project. And finally, some social benefits and costs do not have a market price, or the actual market price or cost may differ from the social value, so that accounting or "shadow prices" must be assigned to them in social accounting.[5]

An example of the first difference is the increase in domestic costs for labor, materials, and the use of capital required to produce a commodity domestically when imports are restricted over the cost of importing the same commodity at the world price. This involves a social cost, but the increased profits to domestic producers due to the higher prices charged to domestic consumers are not a social cost since they simply represent a transfer between domestic parties, not a claim on national physical and human resources. (Chaps. 5 and 7 discuss this matter in the context of supply disruptions of imported minerals.)

An example of the second difference is the pollution generated by a domestic smelter, the social costs of which are borne by others and not by the smelter owners. An external social benefit might take the form of a new technology developed by one producer that became available to others, thus lowering prices to consumers. A final example is the social benefit to the economy resulting from the development of a substitute for an imported material, thereby reducing the vulnerability of the economy as a whole to foreign supply disruption.

An example of the last difference is the monetary value that may be assigned to the social cost of environmental pollution, or to the loss of outdoor amenities resulting from mining in a wilderness area. The social cost of environmental pollution may be estimated in part by the monetary cost of illness and the impairment of human productivity. Although some social scientists reject any attempt to assign a monetary value to human suffering, society implicitly assigns such values through its representatives in government. There are also methods of estimating the monetary value of wilderness amenities (these problems are discussed at some length in chap. 8).

A major contention of this study is that we should abandon the illusory pursuit of a national materials policy and prepare instead specific mineral programs, basing those programs on intensive research and analysis that can best serve this nation's basic interests. Clearly, the financial welfare of the U.S. mining and mineral-processing industries does not qualify in and of itself as a national policy goal, any more than profits and employment in any other single industry constitute a bona fide national goal. What is at question is the economic and social welfare and security of the nation as a whole. Each of the many materials issues is susceptible to optional remedies that affect the promotion of major national policies and goals. In some cases alternative courses of action would have differing effects on competing national priorities such as defense or maximum civilian production, so that choices must be made in such a way as to maximize national welfare as a whole. In other cases, such as those involving governmental action versus private enterprise or restricted versus free international trade, the choice must spring from the principles embodied in the government's own economic philosophy. Special private business interests should play no role in these decisions.

2

Will We Run Out of Nonfuel Mineral Resources?

The basic question posed by this chapter is not actually whether a given mineral will be used up, but rather when world economic growth will be substantially curtailed by rises in the real prices of specific nonfuel minerals.

The world's mineral resources will not run out. Several minerals constitute a high percentage of the earth's crust. Simply projecting present consumption in relation to the volume of known and probable mineral reserves is inadequate to determine the date of exhaustion or severe shortage. Reserves are not static. One hundred years ago deposits containing 0.3 percent copper would not have been classed as reserves, but they have been profitably mined during the past decade. Discovery of more mineral reserves and technological advances in mining and metallurgy make it possible to exploit lower-grade deposits. Technological progress also makes it possible to substitute more abundant minerals for relatively scarce ones, thus conserving scarce minerals without sacrifice of end-product quality or increased real cost.

Conceivably, future technology will make possible substitution of the most common materials for the much rarer ones: aluminum from clay rather than from bauxite, or a variety of minerals from seawater, not from ore. Such substitutions would require enormous amounts of energy, but almost limitless energy is available from the sun.

What is conceivable, however, is not necessarily practical or even technically possible in the foreseeable future. It is impossible to predict now what technology will achieve in discovering substitutes for scarce minerals. Nor is it now possible to know what finding and extracting the almost unlimited supplies of some minerals in the earth's crust and in seawater will cost. The whole process might absorb too much of the world's energy and capital needed for sustained overall economic growth.

For these reasons it is impossible to predict whether and when mineral scarcity will seriously limit world economic growth.[1] It is possible, however, to limit our horizon to specific time periods over which projections can be made and to calculate their probabilities based on what is known today. Such estimations would be based on the expected rate of growth and composition of world output; the rate of growth of demand for minerals required to sustain that rate of output growth; the volume of known and probable mineral reserves and the probabilities of adding to these reserves; and a projection of technological advances likely to permit the substitution of more abundant for less abundant minerals.

Potentially Recoverable Mineral Resources

In dealing with the future availability of nonfuel minerals, it is first useful to identify some basic classifications of resources employed by geologists and mineral economists. A mineral *resource* is a concentration of a naturally occurring material, the commercial extraction of which is presently or potentially feasible. Mineral *reserves* on the other hand constitute that portion of total mineral resources that it is presently economical to extract. The determination of mineral reserves requires estimating the cost of extraction, given present technology, and evaluating the economic feasibility of extracting the minerals at various prices while earning some minimum rate of return on invested capital.

Reserves are classified according to how much exploratory work has been done on them. These classifications include *measured* (or proven) *reserves, indicated* (or probable) *reserves,* and *inferred* (or possible) *reserves.* The determination of measured reserves in an ore body requires sufficient drilling and collecting of core samples to estimate the volume and grade of the ore with a high degree of confidence. This is a costly undertaking, ordinarily limited to deciding on the feasibility of a potential mine. Indicated reserves are determined after less intensive examination. They may be far larger than measured reserves in a given ore body since extensive drilling is costly and companies may not undertake drilling of a deposit beyond that required for a feasibility study. Also, an ore body that is being mined may not be explored to determine measured reserves beyond what is necessary to plan the operation of the mine for several years in advance. However, knowledge of the dimensions and geological characteristics of the ore body make it possible to estimate "indicated" reserves for the total ore body even without extensive drilling. "Inferred" reserves are based on still less knowledge derived from geological evidence.

Identified resources include *reserves* that are presently economic to produce; *marginal reserves,* the economic character of which is uncertain; and *subeconomic resources* that may potentially become reserves (see fig. 2-1). Each of these classifications applies to *measured, indicated,* and *inferred* reserves (or subeconomic resources) based on the amount of exploration work done on them, as described in the preceding paragraph. Hundreds of millions of tons of seabed nodules containing cobalt, copper, nickel, and manganese have been identified as resources, but they are not regarded as reserves because the economic feasibility of their extraction is uncertain. In addition to identified resources, there are two classifications of *undiscovered mineral resources* that potentially may be economically exploited: (*a*) *hypothetical resources* that potentially may be economically exploited and (*b*) *speculative resources* in nonproducing districts but in favorable geologic settings. The identification of hypothetical and speculative resources capable of commercial development is based on a variety of geologic information. However, the vast bulk of undiscovered resources may well lie outside these categories. Although by definition they are potentially recoverable, they are currently subeconomic (see fig. 2-1).

The total volume of any mineral in or on the earth's crust is likely to be very large, even in the case of minerals such as gold or silver, which constitute an exceedingly small percentage of the crust's volume. To be regarded as a resource a mineral must be highly concentrated, in some cases thousands of times more so than its average crustal occurrence. For example, it would require 285 million metric tons (mt) of average crustal rock to produce 1 mt of gold, but in order for gold to be profitably mined, the minimum ratio of gold to ore is 1 mt of gold to 100,000 mt of ore.[2]

Although geologists continually discover more of these concentrations, a question arises regarding the rate at which the volume of reserves is likely to increase with additional geologic knowledge and exploration. Individual ore bodies become progressively exhausted through mining, often ending up with ore grades too low for commercial exploitation. But increased prices of the mineral and improved technology may render the remaining ore commercially exploitable sometime in the future. As more resources are identified by geologists, whether classified as reserve grade or not, a larger proportion of these new discoveries tends to be of lower grade or more difficult to extract. Most of the relatively rich mineral deposits lying at or near the surface have already been discovered.

A general rule of most geologists is that the total *volume* of an extract-

Identified Resources			Undiscovered Resources	
Measured	Indicated	Inferred	Hypothetical	Speculative
Reserves		Reserves		
Marginal reserves		Marginal reserves		
Subeconomic resources		Subeconomic resources		

Fig. 2-1. Major elements of mineral resource classification. (*Source:* Bureau of Mines, *Mineral Commodity Summaries 1982* [Washington, D.C.: U.S. Department of Interior, 1982].)

able mineral increases progressively as the grade of the ore declines. While the total volume of copper found in all ores containing 5 percent or better copper is relatively small, the volume of copper in all ores with a grade of 2 to 5 percent is several times larger, and the volume of copper in ores containing less than 2 percent copper is many times larger. Also, as technological and economic conditions change to permit the economical exploitation of lower grades, the global volume of extractable copper tends to increase.

However, this general rule may not hold for some minerals below a critical ore grade. B. J. Skinner has suggested that the above rule is probably true for minerals such as aluminum and iron, which are quite abundant in the earth's crust, but that there exist quite different distributions of elements such as copper, cobalt, nickel, gold, and silver, which are relatively scarce in the earth's crust.[3] In these cases, the initial period of exploration and mining will be characterized by a rising global volume of the extractable mineral as the grade declines, but below a critical grade the global volume will decline.[4] Insufficient geological information is available to prove Skinner's hypothesis or to determine the grades of various minerals below which the global volume of a given mineral declines. But if Skinner is right, changes in technological and economic conditions are limited in their ability to facilitate mining lower and lower grades of ore to provide increasing volume.

Another hypothesis suggesting a limit to the availability of mineral reserve is that scarce metals occur in proportion to their crustal abundance. Estimates of the total volume of particular metals in the earth's crust concentrated in ore bodies with grades above certain percentages indicate that no more than 0.01 percent of all copper in the continental crust will be found in ore bodies with grades of 0.1 percent copper or more, and that the amount so concentrated may be as low as 0.001 percent of the total in the earth's crust.[5]

Although technological progress in mining and metallurgy has made possible the commercial exploitation of increasingly lower grades, energy requirements increase as grades decline. Energy requirements also differ with specific kinds of minerals so that, for example, silicate minerals require substantially more energy for processing than other forms. The increasing cost of energy has led to the suggestion that energy requirements will pose severe limitations to the exploitation of lower grades and more complex types of minerals.[6] This is likely to be the case unless present energy requirements are drastically reduced by future technology.

Even if we calculate the potentially recoverable resources of a mineral in the earth's crust, we are faced with the question of what proportion of these resources is likely to be found in locations and quantities that render recovery economically feasible. The U.S. Geological Survey (USGS) has estimated that there are 3.4×10^6 mt of recoverable mercury resources at present extraction costs within 1 km of the earth's surface. In 1981 total world production was over 7,000 mt, so that there is enough mercury in the earth to last over five thousand years at the 1981 rate of production. However, the USGS estimate includes all mercury deposits under the ocean (70 percent of the earth's surface), under the Antarctic and Arctic regions, and under mountain ranges and other areas where discovery is highly unlikely. Obviously only a small fraction of these resources is likely to be found; in any case the volume of many concentrations would be too small to mine profitably.[7]

In contrast to the USGS estimate, the U.S. Bureau of Mines (BOM) estimates world mercury resources in accordance with the categories shown in figure 2-1 at about 590,000 mt, or an amount sufficient to support some eighty years' production at current rates.[8] Therefore, in discussing the availability of mineral resources for meeting future requirements, it is more realistic to employ estimates of world reserves and world resources as defined by the BOM than to employ those of the USGS. This is particularly the case when we consider the availability of minerals in finite periods up to, for example, 2050. Beyond this we can do little to project how fast reserves will be discovered, how fast demand for particular minerals will grow, or what technological developments will have important effects on both demand and supply. For example, by the year 3000 technological developments may enable us to substitute those almost unlimited minerals in the earth's crust, such as aluminum (extracted directly from some clays) and iron, for almost any other mineral now essential for industrial production.

Measuring the Scarcity of Minerals

What does it mean to say that nonfuel minerals are becoming increasingly scarce? Scarcity is sometimes used loosely to mean a shortage of supply of a material relative to the demand for it. But in a free market—one in which demand is governed by price—there will always be enough supply to meet demand. Because the price of a nonrenewable resource rises as its supply declines, the total reserves of any mineral are never fully exhausted, no matter how fast demand for it grows. However, relative scarcity is an important economic consideration for nonrenewable resources, so we require a reliable method to measure it.

The two principal measures of relative economic scarcity of a resource are (*a*) the index of real unit cost (i.e., the change in unit cost relative to other prices from a base period) and (*b*) the index of real price (i.e., the change in price relative to other prices from a base period). Both have been used in discussing past and projected trends in the scarcity of nonrenewable resources and each has certain advantages and disadvantages. Over the long run, the real cost of a mineral in terms of the labor and capital required to find and produce it will tend to move parallel with its real price.[9] As the real price of a mineral rises, it will induce firms to incur higher costs of exploration and production until the marginal cost of the additional output is equal to the higher price. For example, production costs of newly discovered reserves in more remote locations and deeper in the ground tend to be higher than production costs in more accessible areas. Therefore, as long as output can be expanded by incurring additional costs, it makes little difference whether we measure relative scarcity by movements of real prices or of real costs.

The traditional view of nonrenewable resources is that economic scarcity increases with rising world consumption. This view was challenged by H. G. Barnett and Chandler Morse.[10] They used the real unit cost measure of scarcity and found that the real unit cost index of minerals declined sharply between 1870–1900 and 1957. The index was 210 in 1870–1900 (1929 = 100) and 47 in 1957. Barnett and Morse also constructed a relative real unit cost index to measure the change in labor plus capital costs of minerals relative to the change in labor plus capital costs in the nonextractive sector. Their relative real unit cost index for minerals was 154 in 1870–1900 (1929 = 100) and 68 in 1957.[11] Extension of the two indexes to 1970 shows a continuation of this downward trend beyond 1957.[12]

Orris Herfindahl found that the real cost of copper declined by about 40 percent immediately after World War I compared to the 1885–1914 period and then remained fairly stable between the early 1920s and 1957.[13] The substantial decline in the real cost of copper following World War I was largely a consequence of rapid technological improvements in the mining and processing of copper in the second decade of the 1900s. Opposing developments maintained the real cost of copper during the period from the early 1920s to 1957. First, the grade of ore mined in the U.S. declined from an average of 1.6 percent (1921–30) to 0.8 (1951–56). The declining grade of ore was balanced by (*a*) technological advances and scale economies, (*b*) a slower rate of growth, both in production and in demand, partly due to competition from scrap and substitute materials such as aluminum, and (*c*) a substantial expansion of copper discoveries and output in regions outside those producing in 1920.[14]

These same opposing forces noted above to explain the movement of real costs (and real prices) of copper over the past century have operated with respect to a number of other minerals, including aluminum, iron ore, nickel, and zinc. They help to explain why the depletion of high-grade ore bodies in countries first producing some metals has not been accompanied by a long-run increase in their real price.

Some experts caution that current real prices may not signal the impending exhaustion of a resource because the market fails to anticipate future conditions.[15] For example, oil selling at $2 per barrel in 1970—in the face of an 8 percent annual growth in world demand for it—was certainly a misleading measure of the scarcity of petroleum. When the OPEC cartel raised the price of its oil in 1973–74, it claimed to be charging the proper rate for the scarce resources it sold. OPEC argues that even without its action, petroleum prices would have risen sharply to reflect the depletion of world reserves.

The underpricing of oil before 1973, however, should not be attributed to a failure of the market to anticipate future prices based on long-term demand and supply estimates. In fact, this "market failure" arose from the oligopolistic structure of a world market dominated by multinational oil companies before 1972. Had the price of oil risen slowly over several decades (i.e., had a free market for the commodity existed), the world would have had time to lower the rate at which petroleum consumption was growing and to undertake more intensive exploration for new reserves. A gradual price increase based on competitive demand and supply conditons (including the gradual depletion of resources) would have been much less harmful to the world economy than OPEC's tenfold upward manipulation in less than a decade. World market conditions in nonfuel minerals are not the same as those that have existed in the petroleum industry: no analogous sudden rise in price is likely for any of them.

Can scarcity be defined without regard to how substitution and conservation can reduce demand as real unit costs and prices rise? No, because scarcity is a function of demand, and in the absence of demand nothing is scarce, even though very little of it may exist. Known world resources of tin are small compared with bauxite and nonbauxite resources for producing aluminum. Yet no one is concerned about skyrocketing tin prices as its supply resources run out, because there are adequate substitutes for most uses of tin. New technology constantly reduces the demand for certain materials and is itself frequently developed in response to a rise in the price of a material. For example, the invention of nitrogen fixation from the air destroyed most of the value of Chile's major export—its nitrate deposits—in the early 1920s. Thus within the dynamic context of technological progress, it may be argued that no mineral is subject to growing scarcity over a long period of time.[16]

Conclusions on Measuring Scarcity

Increasing or decreasing scarcity of a mineral may be measured by long-run trends in either its real price or the real cost of producing it. Real prices will tend to reflect the (real) marginal cost of finding and extracting new reserves. But the development of less costly substitutes for a mineral reduces its price and, hence, its scarcity. This causes marginal costs to fall because the market price does not justify more intensive exploration or the mining of lower-grade and less accessible ore bodies. Only in the rare case of fixed world reserves of a mineral controlled by a monopolist will prices be independent of marginal costs of discovering and extracting additional reserves.

Because it is often difficult to measure the real cost of producing a mineral, long-run trends in real prices are probably the best way to measure economic scarcity. However, market forces do bring about considerable short-run price fluctuations. And as the petroleum example demonstrates, both imperfect competition and the market's failure to reflect future demand and supply relations may result in real prices that understate or overstate economic scarcity for a considerable period of time. Given the advantage of hindsight, it is obvious that $2 per barrel oil did not reflect a long-term petroleum glut in 1970. Enough information is available today to show that $0.59 per pound copper in 1984 understates its long-term scarcity by perhaps a factor of two. Why don't speculators buy copper and other nonrenewable resources when they are obviously undervalued, thereby raising prices? One reason is that the annual cost of storing a commodity (including interest on capital) may amount to over 20 percent of its value.[17] The second reason is uncertainty about long-run growth in demand and the outlook for discovery and development of additional reserves. More information and a better methodology for projecting long-run demand and better geological estimates of existing and probable reserves of minerals could help reduce these uncertainties. But technological advances that may affect both demand and economically recoverable resources are exceedingly difficult to predict over periods of time as long as twenty-five years.

Adequacy of World Reserves and World Resources for Meeting Projected Consumption of Minerals to 2000

The usual method of determining the adequacy of world reserves or resources of a mineral over a finite period in the future is to calculate the life expectancy of the reserves or resources in relation to the projected growth rate in world demand for the mineral. Table 2-1 shows the life expectancy of the world reserve base of twenty-two minerals for four hypothetical annual rates of growth in consumption over the 1978–2000 period, together with probable average annual rates of growth in consumption as projected by the BOM.[18] Except for industrial diamonds, gold, mercury, and silver, the reserve base is ample for projected world consumption well beyond the year 2000. In the case of a number of minerals, including bauxite, chromium, cobalt, iron ore, manganese, and the platinum-group metals, the current reserve base will be adequate for more than twenty years beyond the year 2000. There are also large monetary stocks of gold and silver available for possible industrial use.

It is important to note that in most cases the projected average annual rates of consumption for the 1978–2000 period in table 2-1 are likely to be on the high side. Recent consumption growth projections for bauxite, copper, iron ore, nickel, and tin made by the World Bank and other investigators are

TABLE 2-1. World Mine Production, Reserve Base, Life Expectancy, and Probable Average Annual Consumption Growth Rates for Selected Minerals

Mineral	Unit	1980 Mine Production[a]	Reserve Base[a]	Life Expectancy of Reserve Base at Four Annual Growth Rates (in years)				Probable Average Annual Growth Rate 1978–2000[b] (in percentage)	
				0%	2%	4%	5%	U.S.	World
Bauxite	thousand mt	89,933	22,400,000	249	90	61	53	4.0	3.7–5.2[c]
Chromium	thousand st	10,725	3,700,000	345	106	69	59	3.2	3.3
Cobalt	st	32,965	3,400,000	103	57	42	37	2.5	2.8
Columbium	thousand lbs	32,300	7,600,000	235	88	60	52	6.1	6.1
Copper	thousand mt	7,630	505,000	66	43	33	30	2.4	2.7–3.6[c]
Diamonds (industrial)	million carats	30.9	620	20	17	15	14	4.8	4.8
Gold	million troy ozs	39.0	1,200	31	24	20	19	2.4	1.5
Ilmenite	thousand st	5,363	727,000	136	66	47	42	4.8[d]	5.0[d]
Iron ore	million lt	873.6	108,000	124	63	46	40	1.4	2.3–2.6[c]
Lead	thousand mt	3,520	165,000	47	33	27	25	1.2	2.9
Manganese	thousand st	29,000 est.	5,400,000	186	78	54	48	1.4	2.7
Mercury	76-lb flasks	191,100 est.	4,400,000	23	19	17	16	−0.4	2.2
Molybdenum	thousand lbs	239,190 est.	21,700,000	91	52	39	35	4.2	4.5
Nickel	st	850,366	59,800,000	70	44	34	31	3.7	2.3–3.9[c]
Platinum group	thousand troy ozs	6,830	1,195,000	175	76	62	47	2.2	2.4
Rutile	thousand st	467	145,200	311	100	66	57	4.8[d]	5.0[d]
Silver	million troy ozs	342.2	8,435	25	20	17	16	2.9	1.9
Tantalum	thousand lbs	978	48,000	49	34	27	25	4.1	4.1
Tin	mt	246,250	10,000,000	41	30	25	23	0.9	0.7–0.9[c]
Tungsten	thousand lbs	119,320	6,400,000	54	37	29	27	4.5	3.5
Vanadium	thousand lbs	79,100	40,800,000	516	124	78	67	3.9	3.6
Zinc	thousand mt	5,745	240,000	42	32	25	23	1.6	2.0–2.9[c]

Notes: Reserve base as defined by the Bureau of Mines includes those resources that are currently economic (reserves), marginally economic (marginal reserves), and some that are currently subeconomic (subeconomic reserves) that have a reasonable potential for becoming economically available within planning horizons beyond those that assume proven technology and current economics. mt = metric tons; st = short tons; lt = long tons.

[a]Bureau of Mines, *Mineral Commodity Summaries 1982* (Washington, D.C.: U.S. Department of Interior, 1982).

[b]Bureau of Mines, *Mineral Facts and Problems 1980* (Washington, D.C.: U.S. Government Printing Office, 1981). Annual growth rates refer to growth in the demand for primary metal. For bauxite, annual rate of growth refers to aluminum; for ilmenite and rutile, annual rates of growth refer to titanium.

[c]World Bank, *Price Prospects for Major Primary Commodities,* vol. 1 (Washington, D.C.: World Bank, July 1982), p. 60. World Bank annual growth rates refer to period 1980–95.

[d]projected annual rates of growth for titanium

below those of BOM. Consumption growth rates for most minerals have not only declined substantially from the 1960s and early 1970s, but projected rates to the year 2000 are continually being revised downward.

The reserve base of nearly all minerals has been expanding. For example, over the past decade reserves of bauxite and copper have risen by some 40 percent. Exploration moves more "indicated" or "inferred" resources into the "measured" or "proven" category, and expanded geological knowledge causes the experts to classify a portion of hypothetical or speculative resources as "identified" resources (see fig. 2-1). Increased prices for individual minerals will also move some identified resources from the marginal or subeconomic categories into the reserve category. Therefore, in estimating the adequacy of mineral resources over a period exceeding twenty years, it is appropriate to base calculations on world resources rather than on the reserve base given in table 2-1.

TABLE 2-2. World Resources of Selected Minerals

Mineral	Volume
Bauxite	40–50 billion mt
Chromium	36 billion st gross weight chromite
Columbium	38 billion lbs
Cobalt	6 million st, land, plus 250 million st, sea
Copper	1,627 million mt, land, plus 690 million mt, sea
Diamonds (industrial)	1+ billion carats[a]
Gold	2 billion ozs, of which 15–20% are by-product resources
Ilmenite	1 billion st of contained titanium dioxide
Iron ore	260+ billion st of Fe
Lead	1.4 billion mt
Manganese	3.1 billion st, land, plus 15.0 billion st, sea
Mercury	17 million 76-lb flasks
Molybdenum	46 billion lbs
Nickel	228 million st, land, plus 430 million st, sea
Platinum group	3.2 billion troy ozs
Rutile	280 million st of contained titanium dioxide
Silver	25 billion troy ozs
Tantalum	560 million st
Tin	37 million mt
Tungsten	14.9 billion lbs
Vanadium	120+ billion lbs
Zinc	1.8 billion mt identified plus 2.5–3.0 billion mt hypothetical and speculative

Sources: Bureau of Mines, *Mineral Commodity Summaries 1982* (Washington, D.C.: U.S. Department of Interior, 1982) and *Mineral Facts and Problems 1980* (Washington, D.C.: U.S. Government Printing Office, 1981).

Note: Mineral resources are in such forms and amounts that economic extraction is currently or potentially feasible. The classification of resources applying to each of the minerals listed in this table is not uniform. In most cases they include all identified resources, but may also include hypothetical and speculative deposits (see fig. 2-1).

[a]includes newly discovered Australian reserves estimated as high as 500 million carats

The world resource estimates of twenty-two minerals given in table 2-2 include, in addition to the reserve base, inferred reserves and hypothetical and speculative resources. The estimates are based on geological evidence for resources currently or potentially susceptible to profitable extraction. For most minerals listed, world resources are several times the current reserve base. Moreover, as in the case of the reserve base, continuing geological investigation and exploration will undoubtedly enlarge world resource estimates for all minerals.

Table 2-3 gives the life expectancy beyond 1980 of world resources for the minerals listed in table 2-2, assuming the projected annual rates of growth in consumption for the 1978–2000 period continue into the twenty-first century. Note that for seven of the twenty-two minerals the expected life exceeds one hundred years; for another eight the resources would be sufficient to meet requirements until 2050; and except for industrial diamonds and gold, resources for the remainder would be adequate until the year 2025. Given the likelihood that world resources of these minerals will continue to expand, this analysis should provide a degree of comfort. The only problem is that it is not possible to assume that consumption growth rates during the twenty-first century will conform to those projected for the remainder of this century.

Projecting the Demand for Minerals

Projecting the demand for minerals over the next couple of decades is fraught with uncertainties, and organizations with many mineral specialists—such as the World Bank and the U.S. Bureau of Mines—often disagree substantially (see table 2-4). Projections based on past relationships between consumption

TABLE 2-3. Life Expectancy beyond 1980 of World Resources of Selected Minerals at Probable Average Annual Rates of Growth in Consumption for the Period 1978–2000

Mineral	Years	Mineral	Years
Bauxite	74	Mercury	50
Chromium	Over 100	Molybdenum	52
Cobalt	Over 100	Nickel	Over 100
Columbium	70	Platinum group	Over 100
Copper	75	Rutile	70
Diamonds (industrial)	20	Silver	45
Gold	35	Tantalum	80
Ilmenite	48	Tin	100
Iron ore	80	Tungsten	49
Lead	85	Vanadium	Over 100
Manganese	Over 100	Zinc	Over 100

Sources: See table 2-2 for estimates of world resources. Includes seabed resources for cobalt, copper, manganese, and nickel. For probable average annual consumption growth rates for the period 1978–2000 see table 2-1.

of a material and gross national product (GNP) may prove far from the mark because of the shifting composition of GNP or because of changes in the material composition of commodities for which nonfuel minerals are presently essential. Highly sophisticated econometric models are often used to estimate long-run demand, but they rely heavily on past experience to tie demand for minerals to other variables. Accurate projection is likely to depend upon correct judgments about future technological developments and shifts in demand for products that have used the mineral in the past. For example, new processes as well as the need for certain qualities in metals used to build automobile and aircraft engines may result in a sharp fluctuation of demand for alloyed metals like cobalt, manganese, and nickel. Relative price changes, such as the comparative expense of tin and that of substitute materials for beverage containers, may also affect rates of growth in demand for minerals. Because long-run estimates have to be made without adequate knowledge of future developments, a range of demand projections to which probability coefficients are assigned is likely to be more suitable for private investment or public policy decisions than the use of single projections.

Consumption growth rates for most minerals have been declining in recent years as the composition of GNP has shifted toward services (particularly in developed countries) and as technological progress has made for less

TABLE 2-4. Historical and Projected Rates of Growth in World Consumption of Selected Nonfuel Minerals (in percentage per annum)

		Projected			
Mineral	Historical	Fischman[a] 1980–2000	World Bank[b] 1980–95	BOM[c] 1978–2000	Malenbaum[d] 1971–75/2000
Aluminum	7.0 (1961–80)[a]	3.7	3.7	5.4	4.1
Bauxite	6.0 (1964–77)[b]	4.0	3.6	5.2	n.a.
Chromite	4.5 (1960–76)[b]	2.8	n.a.	3.3	3.0
Cobalt	5.0 (1960–76)[b]	1.8	n.a.	2.8	3.3
Copper	3.4 (1961–80)[a]	2.5	2.7	3.6	2.8
Iron ore	3.3 (1961–80)[a]	n.a.	2.3	2.6	2.8
Lead	3.5 (1961–80)[a]	2.0	3.1	2.9	n.a.
Manganese	4.0 (1961–80)[a]	2.6	2.6	2.7	3.1
Nickel	4.5 (1969–79)[c]	n.a.	2.3	3.9	2.8
Tin	0.2 (1961–80)[a]	n.a.	0.7	0.9	1.9
Zinc	3.5 (1961–80)[a]	2.6	2.9	2.1	2.9

Note: n.a. = not available

[a]L. L. Fischman, *World Mineral Trends and U.S. Supply Problems* (Washington, D.C.: Resources for the Future, Inc., 1980).

[b]World Bank, *Price Prospects for Major Primary Commodities,* vol. 1 (Washington, D.C.: World Bank, July 1982), p. 58.

[c]Mine production: Bureau of Mines, *Mineral Facts and Problems 1980,* (Washington, D.C.: U.S. Government Printing Office, 1981).

[d]Wilfred Malenbaum, "Anatomy of World Materials Demand: Its Role in Supply Availability," *Mining Congress Journal,* February 1979.

intensive use in particular products. Moreover, future growth rates in the demand for most minerals are projected to decline from those of the past two decades. For example, world consumption growth rates for copper and aluminum were 3.4 percent and 7.0 percent per annum respectively in 1961–80. By contrast, the World Bank projects annual growth rates of 2.7 and 3.7 percent for copper and aluminum, respectively, for the 1980–95 period.[19] As table 2-4 shows, for every mineral except tin, projected world consumption growth rates are substantially below their historical counterparts. Moreover, for a number of other minerals not listed in table 2-4, including industrial diamonds, molybdenum, the platinum-group metals, and vanadium, anticipated rates of consumption growth are also below historical norms.[20] The reasons are again the shift in composition of GNP in the developed countries toward services and the less intensive use of metals in industrial production.

World Demand for Nonfuel Minerals in the Twenty-first Century

Leontief's Projections to 2030

A recent study by Wassily Leontief and associates makes one of the most ambitious efforts so far to predict relationships between world demand and resource availability of major nonfuel minerals in the next century.[21] This study employs a world input-output model that summarizes the flow of goods and services among all producing and consuming sectors of the world economy. The resource input structure of each producing sector depends on existing technology described in the model in terms of the amounts of each input, including nonfuel minerals, required by a sector for each unit it produces. The world demand for each of twenty-six nonfuel minerals is computed under alternative assumptions about income and population growth rates for each of several world regions. Production technologies for final products are assumed to be those anticipated for the year 2000.[22]

The Leontief study makes world demand and supply availability projections for 2010, 2020, 2030. Optimistic as well as pessimistic gross domestic product (GDP) growth rates are analyzed, both for the developed countries and for each of several developing-country regions over the period 2000–2030.[23] The study encompasses all world economies, including the Soviet Union, in its demand projections. The resource base for determining mineral availabilities consists of economic and subeconomic reserves and hypothetical resources as estimated by the BOM, but in some cases resources are less than those given in table 2-2, in part because Leontief excludes minerals contained in seabed nodules.[24]

This analytical model projects that world resources of iron, molybdenum, nickel, tungsten, chromium, copper, bauxite, vanadium, platinum, and titanium will be sufficient to meet cumulative world demand through 2030, even assuming the more optimistic (higher) rates of growth in GDP. The resource base for lead is expected to be depleted by the year 2020; that for zinc before 2010. World mercury and tin resources should run out before the year 2030 if high GDP growth occurs.[25] Definitive depletion dates are not given for all minerals, but in the case of some, such as manganese and cobalt, successful exploitation of seabed nodules would result in major additions to the resource base. Even for minerals whose exhaustion is anticipated before 2030, the Leontief study suggests that additional resources may be located and that increased recycling, together with the use of substitutes, may avoid shortages. Moreover, technological changes reducing the ratio of mineral input to final product are almost certain after the year 2000.

World Demand after 2030

Leontief's results indicate that world demand for most nonfuel minerals is unlikely to exceed available supplies before 2030.[26] Mineral economists differ as to whether the scarcity of nonfuel minerals will limit economic growth thereafter. Some argue that mineral resource depletion is inevitable sometime in the next century, given continued growth in world per capita GDP and population. The basic argument goes something like this.

In 1980 the industrial countries (including the Soviet bloc), with less than 25 percent of the world's population, consumed at least 80 percent of the nonfuel minerals used. In the next century the developing countries' demand for these minerals will increase significantly. (In the coming decades it is assumed that rates of growth of both GNP and population will be higher in developing than in developed countries; this has been the case since 1960.) Most developed countries will have zero population growth by early in the next century, while most developing countries will not reach that stage until late in the twenty-first or even in the twenty-second century.[27]

During the 1970s the average per capita GNP of the developing countries grew at an annual rate of 3.1 percent (in constant 1980 dollars) to an average level of $730 in 1980.[28] Projecting this rate of growth to the year 2050 would raise the average per capita GNP of the developing countries to nearly $6,000—about two-thirds of that of the industrial countries in 1980. By 2050 world population should be about double what it was in 1980, reaching 9.8 billion.[29]

Let us assume that in 2050 world annual per capita GNP is $6,000 in 1980 dollars. If the same ratio of world nonfuel mineral consumption to GNP prevails in 2050 that existed in 1980 for the industrial countries plus the Soviet bloc, world mineral consumption would rise by a factor of 4.4. This would mean a cumulative resource consumption of 150 times that of 1980.

Without increase in the 1980 world resource base, five of the minerals listed in table 2-3 (industrial diamonds, gold, mercury, silver, and tin) would be exhausted before 2050. At the rates of consumption growth projected for 1978–2000 in table 2-1, resource exhaustion would occur by 2050 or shortly thereafter for bauxite, copper, and iron ore (see table 2-4). And all but a few nonfuel minerals would be exhausted by the end of the twenty-first century.

The conclusion drawn from the hypothetical scenario is unwarranted because it disregards advances in technology, which have been accelerating. There are some minerals, such as aluminum, iron, and silicon, that are superabundant in the earth's crust and may substitute for less abundant materials.[30] Moreover, the potential for recycling and conservation of less abundant minerals is enormous. It may be that technology and energy are the only true growth-limiting factors. We simply do not know when, if ever, the availability of nonfuel minerals will constitute a limit on world economic progress.

Future Real Costs and Prices of Nonfuel Minerals

Future world economic growth implies adequate mineral resources and the technological capacity to extract them. But it also requires that extraction can be done at a real cost compatible with growth: the real cost (in capital and energy) of the necessary minerals could increase so much as to make the cost of extracting and processing them prohibitive.

Reserves of most nonfuel minerals suggest that declining ore grades are unlikely to lead to significant cost increases during this century. There will surely be a decline in the grade of ore produced, particularly in the older mineral-producing countries, but other countries have substantial reserves of higher grade ores.

Copper provides a good example. At the end of 1976 the estimated average grade of U.S. copper ore reserves was 0.62 percent; the world average for producing mines was higher than 1 percent.[31] A larger part of U.S. copper use will probably come in the future from countries (Chile, Peru, Zaire, and Zambia, for example) where ore grades are significantly higher. How rapidly the average yield of world copper ores will decline over the next two decades is hard to predict, but experience suggests that it is likely to be partially offset by technological advances.

The widely held assumption that ore grades in nonfuel mining have rapidly declined over the past half century seems unduly pessimistic. Canada, one of the world's largest producers and exporters of nonfuel minerals during most of this century, is a case in point. Two Canadian Department of Energy, Mines, and Resources geologists conclude in a recent paper that "the average mining grades of nickel, zinc, nonporphyry copper and gold ores show remarkable steadiness over the past half century 1939–1989. . . . The average grades of lead ores, silver ores, porphyry copper ores, and molybdenum ores declined since 1939." The authors point out that in some cases the lowered ore grades resulted more from changes in mining methods than to depletion of higher-grade ores.[32]

True, ore deposits have become more difficult to find by conventional methods, but according to another Canadian government geologist, new exploration technologies have "greatly increased the rate of discovery and lowered discovery costs for various geological types of deposits in Canada over the rate and costs that would have existed without the development of new technology." Moreover, he points out that, despite the rapid increase in Canadian mineral production since World War II, discoveries of new lodes have substantially exceeded production in most years.[33] Finally, it should be noted that in most mineral-rich developing countries, such as Chile, Peru, and Zaire, there are large ore bodies that have not been developed to the extent typical of countries such as the U.S. and Canada.

Reliable historical data on average real costs of producing individual minerals are not available, but the evidence shows that declining ore grades

TABLE 2-5. Weighted Price Index of Selected Metals and Minerals in Constant Dollars, 1950–84 and Projected 1985–95 (1979–81 = 100)

Year	Price Index	Year	Price Index
1950	124	1970	141
1951	133	1971	116
1952	149	1972	106
1953	139	1973	132
1954	136	1974	144
1955	155	1975	113
1956	158	1976	110
1957	140	1977	104
1958	128	1978	92
1959	125	1979	103
1960	125	1980	105
1961	121	1981	92
1962	116	1982	83
1963	115	1983	88
1964	137	1984	84
1965	155		
1966	160	projected	
1967	134	1985	79
1968	138	1986	73
1969	143	1990	78
		1995	82

Source: World Bank, "Half-Yearly Revision of Commodity Price Forecasts," mimeographed (Washington, D.C.: World Bank, January 1986), table 3, p. 11.

Notes: Weighted by 1979–81 developing countries export value. Commodities include copper, tin, nickel, bauxite, aluminum, iron ore, manganese ore, lead, zinc, and phosphate rock.

have not contributed significantly to increases in real costs since World War II. This may hold true until the end of this century.[34] Not only the Canadian experience suggests this, but also the existence of ores in developing countries of higher grade than the average currently mined.

The major factor behind rising costs of mining nonfuel minerals during the 1970s and early 1980s has been the sharp increase in per ton capital costs. Several reasons account for this, such as increased capital intensity (substituting capital for labor); the cost of capital equipment, which has risen faster than manufactured goods generally; and the high rates of interest, which make borrowing more expensive. In the U.S. and in some other countries as well, capital expenditures for pollution control have substantially increased. (Between 1973 and 1979 about 40 percent of U.S. copper firms' capital went to this purpose.) Finally, drastically higher energy prices have resulted in swollen capital and operating costs both for mining and for processing.

Table 2-5 shows that the weighted price indexes in constant dollars of ten major minerals and metals during the 1977–84 period were the lowest in recent history, with a decline of 2 percent projected between 1984 and 1995. The 1995 figure is substantially lower than the index for nearly every year between 1950 and 1977. This is good evidence that the real cost of producing nonfuel minerals has not been rising during the post–World War II period—and is not expected to rise above the 1950–77 average by 1995.

Table 2-6 shows the indexes of average prices for nine metals in the decade 1960–70, and for the years 1976, 1980, 1981, 1982, 1983, and 1995 (projected). Only for aluminum, copper, and zinc are real 1995 prices expected to be higher than in 1981, whereas those for copper, aluminum, nickel,

TABLE 2-6. Indexes of Commodity Prices and Price Projections in 1981 Constant Dollars

Mineral	1960–70 (average)	1976	1980	1981	1982	1983	1995 (projected)
Aluminum[a]	111	86	87	100	102	107	106
Bauxite	79	100	98	100	92	91	92
Copper	201	118	120	100	86	96	103
Iron ore	209	132	101	100	108	103	95
Lead	109	90	119	100	77	61	79
Manganese ore	146	127	89	100	100	95	87
Nickel[b]	85	97	95	100	95	98	77
Tin	73	79	113	100	92	96	89
Zinc	104	124	86	100	90	95	112

Source: Indexes were calculated from price data in World Bank, *Price Prospects for Major Primary Commodities*, vol. 4 (Washington, D.C.: World Bank, September 1984).

[a]U.S. producer list price

[b]Canadian producer price

lead, iron ore, and manganese ore are all projected to be lower in 1995 than their 1960–70 average. (Additional data on long-term trends in real prices of major nonfuel minerals are presented in chap. 4.)

If the above estimates prove correct, average real costs for major nonfuel minerals as a group will not be significantly higher at the end of this century than they were in the 1960–70 period: if a growing scarcity of major nonfuel minerals does occur, it will not be before the next century.[35] (However, real costs of mining do not include all the environmental costs borne by society.[36] These will be discussed in chap. 8.)

Conservation and Technological Advances

Substitution and conservation are important in moderating the demand for minerals. Price changes normally influence demand only after a substantial lag, since in most cases substitute materials for the same manufacturing or processing equipment are not readily available, and new technology may even need to be developed. Nevertheless, in the longer run, price rises are likely to promote substitution or conservation. U.S. tin consumption in most industrial uses declined substantially during the 1970s, when prices for the metal were rising sharply. Much of the decline was due to technological advances that permitted conservation in the use of tin in tin plating and the development of other materials for beverage containers.[37] U.S. demand for platinum greatly increased during the 1970s as a result of its widespread use in catalytic converters, despite a substantial rise in its price. However, engine pollution can be reduced by using a different technology or less platinum, so if its price continues to rise these alternatives will become more likely. To take another example, the sixfold increase in the price of cobalt following the 1978 invasion of Shaba Province (Zaire) resulted in the substitution of other materials for cobalt. The price rise stimulated research to develop new alloys using less cobalt or, in some cases, no cobalt at all.

Given time, substitutes can be developed for virtually all uses of a metal. This applies to rare metals such as cobalt and chromium that are of critical importance in the aerospace industry as well as to more common metals like copper. Optic fibers and silicon microchip satellites are replacing copper in communications just as aluminum replaced copper in long-distance power lines. Optic fibers can be made of glass, the raw materials for which are exceedingly abundant compared to copper; aluminum can be made from clay found in substantial quantities on the earth's surface. Ceramics, also made from clay, have the potential to replace metal in engine blocks. It seems probable, therefore, that technology will substitute more abundant materials for scarce minerals threatened by depletion in the next century.

Technological Advances in Exploration, Mining, and Metallurgy

Technological advances in mining and processing have served to prevent or moderate increases in the real cost of minerals resulting from the depletion of higher-grade, easy-to-find reserves. These improvements have come at several levels. First, important progress has been made in exploration, which no longer depends upon surface occurrences to locate ore bodies. Geochemical exploration, which involves the systematic measurement of one or more chemical properties of a naturally occurring material, enables geologists to discover abnormal chemical patterns that may indicate the presence of mineral deposits worth exploitation. Measurements are made of the concentration of elements found in rock, soil, or streambed sediment, in vegetation in streams and lakes, and in glacial debris. Discovery of the Ok Tedi copper/gold ore deposit in Papua New Guinea was due to analysis of chemical properties in sediments many miles downstream from the ore body itself.

Geophysics also serves the cause of exploration: sophisticated electronic equipment can detect subtle contrasts in such physical properties as specific gravity, electrical conductivity, heat conductivity, seismic velocity, and magnetic susceptibility. Measurements may be made from the air, on the earth's surface, or in boreholes. In recent years "telegeologic" or "remote sensing" techniques have provided information about complex structures in regions where much of the bedrock is concealed. Side-looking radar imagery or photography has provided useful base maps in areas where conventional photographic methods are inadequate. Satellites and the space shuttle have provided information on potential mineral deposits in unexpected areas.

Advances in mining technology and metallurgy during the present century have made it possible to extract minerals profitably from ores containing as little as one-tenth the ratio required by methods prevailing at the end of the nineteenth century. Today it is possible to chemically process complex ores that only a few decades ago would have been useless. Moreover, this has been done with relatively little increase in real cost per unit of the mineral mined. However, if the reserve base is expanded to include the large volume of potentially economic resources, substantial advances in both mining technology and metallurgy must first occur. A recent study prepared by the Committee on Mineral Technology of the National Academy of Sciences concludes that "presently available technology does not provide the capability for meeting the problems anticipated for the mid- to long-term future when oregrades are likely to be lower, and environmental constraints have been made more severe."[38]

Progress is already occurring. A new technique called "*in situ* leaching" permits the recovery of minerals from lower-grade ores with minimal disruption of the surface. In this process, chemical solutions are pumped into the ore deposit in the ground, and metal-bearing solutions are recovered. The metal-bearing solutions are then treated on the surface to recover the metal.

Another method—"borehole mining"—uses high-pressure liquid jets to dig cavities at the bottom of boreholes. The resulting slurries are pumped to the surface for treatment. This makes it possible to recover minerals from economically marginal and submarginal ore bodies, including those that are very deep, without major disruption of the surface. This method has already been used to get at nonmetallic minerals such as sulfur and potash, but its wide use for metals extraction will require advances in mining engineering as well as in metallurgy.

Traditional smelting methods, which separate metal from waste materials and eliminate impurities such as sulfur, create unacceptably high levels of air pollution and use large amounts of energy. The inability of many pyrometallurgical smelters (furnaces operating at high temperatures) to meet Environmental Protection Agency (EPA) emission standards at acceptable costs has led management to close several of them. An alternative method is called hydrometallurgical processing. This technique involves leaching and chemical extraction without the use of furnaces. Hydrometallurgical processing is environmentally pure (except for the waste water that must be treated for safety) and can operate on a smaller scale than pyrometallurgical smelters. Several hydrometallurgical processes for the treatment of copper concentrates were developed during the 1970s, but each of these has certain disadvantages. Considerable research is necessary to improve these techniques and extend their use to metals other than copper.

In recent years there has been a marked decline of technological innovation in the U.S. primary metals industry. In the mid-1970s research and development (R&D) expenditure was only 0.5 percent of sales in the industry, as contrasted with 3.8 percent for machinery, 3.5 percent for electrical equipment, and 5.2 percent for instruments.[39] Moreover, U.S. primary metal firms have been slow to utilize the advanced technology already in existence. Many European and Japanese firms have maintained a higher rate of technological progress than their U.S. counterparts.

A National Academy of Sciences report suggests several reasons for this lag in U.S. technology.[40] One is that the high capital intensity of the U.S. mineral industry discourages the scrapping of old equipment and its replacement with new technology. Low profits (or even losses) in the U.S. mining industry and the heavy capital requirements associated with installing environmental protection equipment have made it difficult to raise the financing needed for more advanced technology. Although in the past several years many U.S. mining companies have been acquired by petroleum and other companies with substantial financial resources, low metal prices have made these companies reluctant to expand investments in mining.

Recycling

Recycled scrap material constitutes a substantial proportion of total consumption for a number of minerals. The individual degree of recycling varies considerably, however, depending upon the cost of collection and the cost of scrap processing relative to the cost of production from ores. Scrap prices tend to vary according to those of primary metals. U.S. consumption of copper, steel, and lead scrap is relatively high. Copper scrap is used both in the production of refined copper and directly (without initial refining) in the production of copper alloy products. Over 40 percent of total U.S. copper metal consumption comes from scrap.[41] During the decade ending in 1979, an estimated 43 to 46 percent of U.S. steel output was derived from iron and steel scrap.[42] In 1981 the recovery of lead from old scrap was 44 percent of U.S. consumption.[43] Aluminum recovery from scrap has been increasing in the United States; in 1981 it accounted for about 15 percent of U.S. aluminum use.[44] Recovery of secondary nickel accounted for 23 percent of total U.S. nickel consumption in 1981.

Knowledge about how much scrap is collected and available for recovery is inadequate for most metals, but the reserve of unrecovered material is definitely rising. For most metals the contribution of secondary materials to total consumption would increase if future cost-price relationships favored production from scrap over production from ores. New technology is required to recover metals from certain types of scrap. New processes have recently been developed for recovery of cobalt and chromium from obsolete products. Although secondary recovery currently supplies only a small part of world consumption of many nonfuel minerals, recycling could be a major factor in holding their costs down as higher-grade ores become depleted.

Conclusions

Scarcity may be measured by changes in either the real unit cost of producing a mineral or by changes in its real price. Over the long run and given reasonably competitive markets, the real price of a mineral will reflect the real marginal cost of finding and extracting new reserves of a mineral. Given the variety of ore grades and conditions of production, changes in real costs are usually difficult to measure. Therefore, changes in real prices provide the best measure of long-run changes in the scarcity of a mineral. Projections of real prices to the end of this century indicate that they will not rise above the levels of the 1950–77 period.

Given the foregoing analysis of world reserves and resources plus increasing possibilities for substitution and conservation, a scarcity of nonfuel minerals great enough to significantly diminish world economic growth is unlikely between now and well into the next century—say to 2030 or beyond. There is little basis for speculation into the second half of the twenty-first century because the course of technological progress for substituting more abundant for less abundant resources or the rate of new discoveries of resources cannot be projected. Moreover, the sources and costs of energy so far into the future are unknown. Given relatively low-cost energy, lower and lower grades of mineral ores can be economically extracted and processed, thus increasing the world reserve base for most minerals substantially. Projected rates of growth in demand for nonfuel minerals between now and the year 2000 show that estimated world resources for most minerals should be sufficient to meet requirements without significant increases in real costs by the year 2000, and without prohibitively large increases in real costs well into the next century.

It seems wise to temper this optimism in view of the prevailing uncertainty about the rate of growth and composition of world GNP in the next century. This uncertainty stems in part from the difficulty of projecting the growth in per capita GNP in the developing countries. These countries consumed less than 20 percent of the nonfuel mineral supplies in 1980, but if by 2050 their per capita GNP were to rise to the average level of the industrial countries in 1980, presently known world reserves of most nonfuel minerals could be exhausted before or shortly after the middle of the twenty-first century. However, even this eventually could be avoided if technological progress permitted the substitution of abundant minerals in the earth's crust for less abundant minerals.

Technological advances have helped limit the rise in real costs of minerals during the twentieth century despite a substantial reduction in the grades of ore mined and the increasing difficulty of finding new mineral reserves. R&D shows promise of reducing world dependence on a number of relatively scarce minerals by means of conservation, recycling, and substitution. A continuation of this process will limit the increase in the real cost of minerals even as world GNP continues to grow.

3

Production, Ownership, and Investment Patterns

The future adequacy of nonfuel mineral supplies for the United States and other non-Communist nations depends not simply upon the global volume of potentially recoverable resources, but on their geographical location, their production facilities, and the global outlook for exploration, mining, and processing. Government mineral policies and the availability of capital and technology will determine in large measure the future rate of growth in mineral output. Consumption of nonfuel minerals is now heavily concentrated in the major industrial countries, but they are increasingly dependent on mineral-surplus developed countries, such as Australia and Canada, and upon the less developed countries. The Soviet Union is both a large consumer and a large producer of nonfuel minerals. Although it has substantial reserves of these minerals and has in the past been an important exporter of them, the USSR's nonfuel mineral exports to Western industrial countries have declined sharply in recent years. Moreover, mineral supplies from the Soviet Union are not a reliable source because their volume is more often a function of political rather than economic factors.

The dispersion of nonfuel mineral resources and producing capacities in relation to the geographical pattern of their consumption indicates that the degree of import dependence will inevitably increase for the United States and other industrial countries. This means the adequacy of nonfuel mineral supply will depend on the policies of mineral surplus countries and on the international flow of capital, technology, and other factors necessary for discovery and production. The industrial nations are not only increasingly dependent on foreign ores, but also on imports of processed minerals such as steel, and nonferrous and alloyed metals.

Although reserves and processing capabilities for most nonfuel minerals were adequate or more to meet requirements in the early 1980s, world growth will require a substantial increase in investment for new exploration and production capacities between now and the end of the century. These large investments must also be made long before the additional capacities are needed. Investment shortfalls are more likely in the mineral-rich LDCs than in the developed mineral-producing countries.

U.S. and Regional Dependence on Nonfuel Minerals

U.S. import dependence on nonfuel minerals is relatively small (about $19 billion) for an economy with a GNP of nearly $3.5 trillion and total merchandise imports of over $250 billion (1983).[1] A more significant measure of U.S. dependence is provided by the *percentage* that imports occupy in our consumption of important nonfuel minerals (see fig. 3-1). For example, bauxite and alumina, for which the United States is 94 percent import-dependent, are the basic raw materials for producing aluminum. Several of the metals—including chromium, manganese, and nickel—for which the United States is two-thirds or more import-dependent are used for alloys essential to a wide range of industrial products. (For the principal industrial uses of major nonfuel minerals, see app. table A3-1.)

In addition to increasing its reliance on foreign unprocessed metals, the United States is also increasing its import dependence on processed metal alloys such as ferromanganese and ferrochromium.[2] The United States imports virtually all its bauxite, and an increasing proportion of its alumina is also imported, mainly from the bauxite-producing countries.[3]

The implications of import dependence extend to America's allies in the European Economic Community (EEC) and Japan, all of whom are more heavily dependent on mineral imports than is the relatively well-endowed United States. For example, both the EEC and Japan are heavily import-dependent for copper and iron ore. By contrast, the USSR is self-sufficient, or nearly so, in most important minerals; the notable exceptions are bauxite and cobalt (see fig. 3-2).

Geographical Distribution of Production and Reserves

The industrial countries of North America, Western Europe, and Japan consume over 80 percent of the nonfuel minerals produced in the non-Communist world, but the largest non-Communist producers of fifteen of the twenty-two minerals listed in appendix table A3-2 are in the developing countries plus South Africa.[4] The major producers of nonfuel minerals among the Western developed countries are Australia, Canada, South Africa, and the United States. The latter is a large producer of, and potentially self-sufficient in, copper, iron ore, and lead, but nevertheless imports a portion of its needs for these minerals, in part because of the abundance of higher-grade ores in other countries. The developed countries produce the largest share of the mine output of titanium (ilmenite and rutile), lead, mercury, molybdenum, nickel, and zinc and are substantial (but not majority) producers of bauxite, copper, iron ore, silver, tantalum, tungsten, and vanadium. Although over half the bauxite is produced in developing countries, the bulk of alumina and alumi-

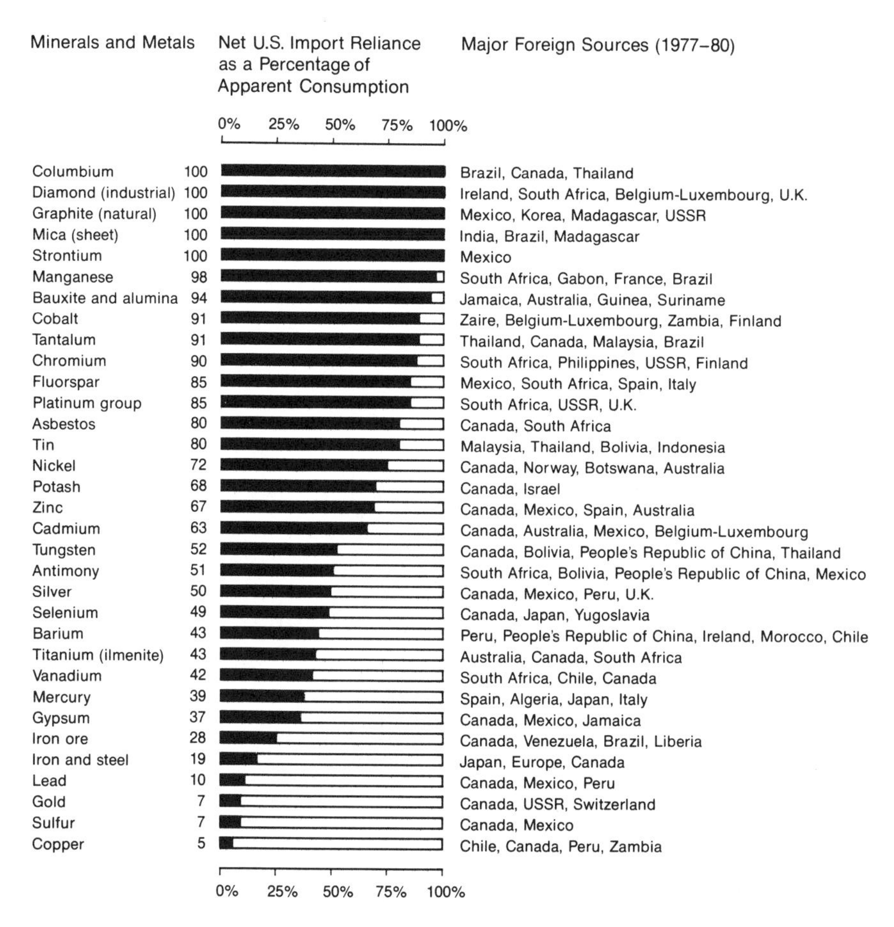

Fig. 3-1. U.S. net import reliance on selected minerals and metals as a percentage of consumption in 1981. Sources shown are points of shipment to the United States and are not necessarily the initial sources of the materials. Net import reliance equals imports minus exports plus adjustments for government and industry stock changes. Apparent consumption equals U.S. primary plus secondary production plus net import reliance. Substantial quantities of rutile, chenium, and zircon are imported, but data are withheld to avoid disclosing proprietary information. (*Source:* Paul R. Portney, ed., *Current Issues in Natural Resource Policy* [Baltimore, Md.: Johns Hopkins University Press for Resources for the Future, 1982], p. 75.)

num are produced in industrial countries. The developed countries also produce most of the refined copper, ferromanganese, steel, and other processed minerals, but a large proportion of the raw materials for these products comes from LDCs plus South Africa. Over the next two decades the relative share of the developing countries in producing raw and processed minerals is projected

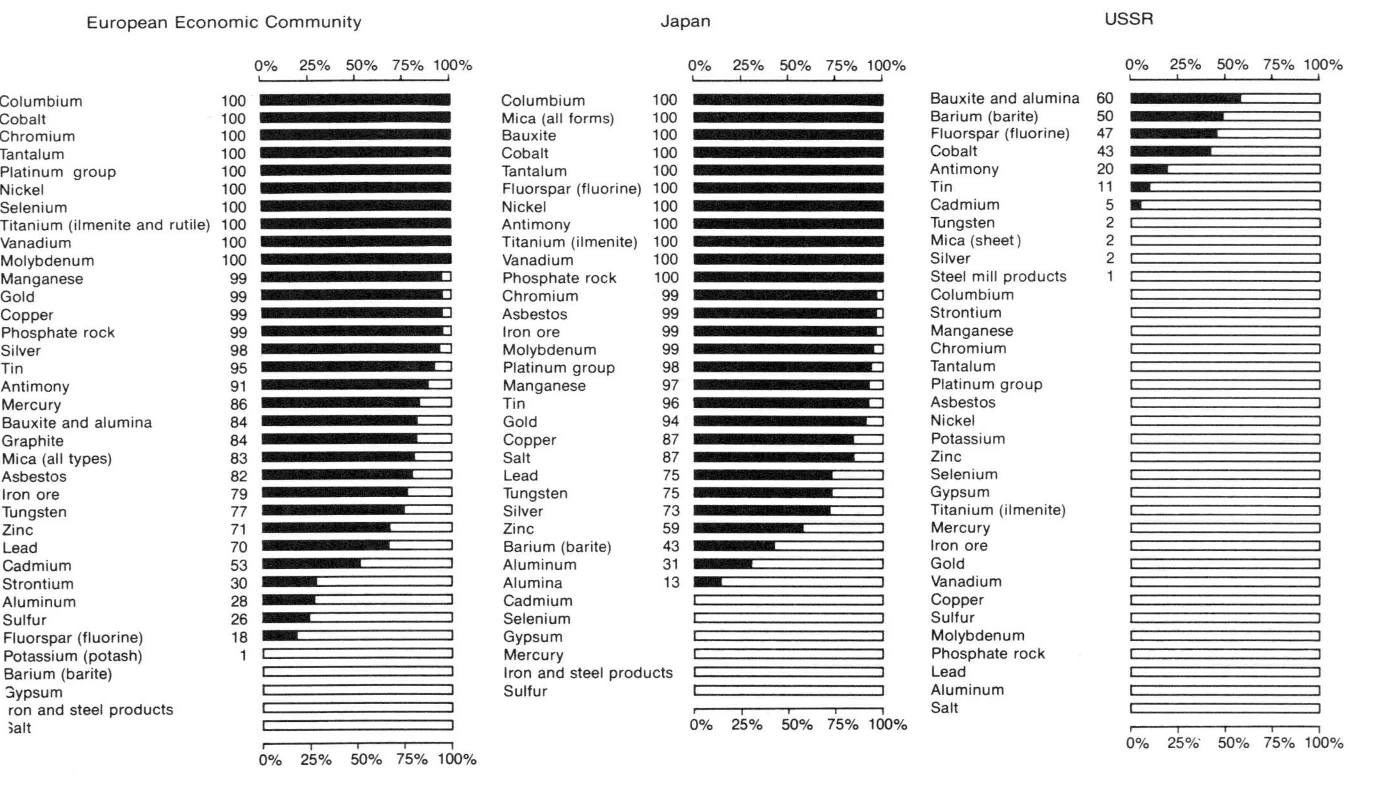

Fig. 3-2. Net import reliance on selected minerals and metals as a percentage of consumption in 1980. European Economic Community in 1980 included Belgium, Denmark, France, Federal Republic of Germany, Ireland, Italy, Luxembourg, the Netherlands, and the United Kingdom. The trade in columbium, tantalum, and vanadium is reported together for the EEC. Data for potassium, strontium, and graphite are not available for Japan. (*Source:* Paul R. Portney, ed., *Current Issues in Natural Resource Policy* [Baltimore, Md.: Johns Hopkins University Press for Resources for the Future, 1982], p. 101.)

to increase. They, along with South Africa, hold a larger share of the most important mineral reserves than do the developed countries. In addition, their share of the world's resources from which additional reserves will be established by increased exploration is substantially larger than that of the developed countries.

The USSR is self-sufficient in most nonfuel minerals and is an important exporter of chromite and the platinum-group metals. During the 1960s and early 1970s it was an important exporter of manganese to the non-Communist world, but during the late 1970s these exports declined sharply; by 1980 virtually all USSR manganese ore was either consumed domestically or exported to Eastern Europe. Soviet exports of chromite ore to the Western world also fell steeply—from an average of about 460,000 mt during the 1971–75 period to less than 100,000 mt in 1980.[5] Exports of platinum also declined by more than 50 percent between 1977 and 1980–82, but since the same production level has been maintained, it appears that the USSR has either increased its domestic consumption or has withheld exports in expectation of higher prices. Although the Soviet Union has large reserves of most major nonfuel minerals, it has had increasing difficulty in meeting production goals in both mining and mineral processing. This has been due to low productivity, outmoded technology, and perhaps poor management.[6]

China has vast mineral resources, as yet inadequately explored. Its deposits of iron, copper, bauxite, lead, zinc, and tin are reported to rank among the largest in the world, and its tungsten reserves are larger than the combined reserves of the rest of the world. China is an important exporter of antimony, mercury, tungsten, and tin, but it imports aluminum, copper, and steel despite its own substantial reserves of the ores for these metals. A recent paper provided evidence that investment in mining and metallurgy in China has declined over the past several years and that China's nonfuel minerals industry is unlikely to develop rapidly.[7]

Distribution of Processing Capacity

The predominant position of the United States in steel, aluminum, and other metals declined substantially between 1950 and 1980 (see table 3-1). In 1950 it produced 50 and 55 percent, respectively, of the world output (excluding China) of aluminum and steel; by 1980 its shares had declined to 25 and 21 percent. The U.S. share of refined zinc output also fell dramatically, from 41 percent in 1950 to only 8 percent in 1980. Since 1950 the production of processed nonfuel minerals has shifted heavily to the LDCs, Japan, South Africa, and the newer industrialized countries (Australia and Canada). The LDCs increased their share of aluminum, steel, and zinc output quite substantially and Japan's position rose from 2 or 3 percent in 1950 to that of a major world producer of refined copper, steel, and zinc. Between 1950 and 1980 Western Europe's share of non-Communist-world aluminum, refined copper, and zinc output increased, although its share of the world output declined. Western Europe's 1980 share of the world output of all these commodities, however, was less than it had been prior to World War II.

These changes in relative shares have been accompanied by important changes in world trade patterns for the same minerals. Data comparable to the international trade matrix given in appendix table A3-3 are not available for earlier than 1973, but important trade shifts are apparent thereafter. For example, the value of U.S. imports (in 1980 prices) of both nonferrous metals and iron and steel has risen severalfold between 1973 and 1980; this is also true for other industrial countries. Exports of both nonferrous metals and iron and steel products from Australia and South Africa, mainly to industrial countries,

TABLE 3-1. Share of World's Mineral Processing Capacity or Output of Selected Minerals, 1950, 1977, 1980, and Projected 1990, in Percentage

Mineral	1950 (output)	1977 (capacity)	1980 (output)	1990 (projected output)
Primary aluminum				
Developing countries	neg.	13	20	25
Developed market economies	99	87	80	75
United States	50	36	35	27
Western Europe	17	28	25	24
Japan	2	14	8	4
Refined copper				
Developing countries	36	27	36	38
Developed market economies	64	73	64	62
United States	37	29	24	n.a.
Western Europe	15	20	20	n.a.
Japan	1	14	14	n.a.
Steel				
Developing countries	4	9	18	32
Developed market economies	96	91	82	68
United States	55	28	21	21
Western Europe	31	35	29	22
Japan	3	23	22	18
Zinc metal				
Developing countries	5	13	23	31
Developed market economies	95	87	77	69
United States	41	11	8	7
Western Europe	30	41	32	30
Japan	3	17	16	15

Sources: Estimated 1950 output: W. S. Woytinsky and E. S. Woytinsky, *World Population and Production: Trends and Outlook* (New York: Twentieth Century Fund, 1953), chaps. 22 and 29. 1977 capacity: United Nations, *Mineral Processing in Developing Countries* (New York: United Nations, 1980), chap. 2. 1980 output and projected 1990 output: World Bank, *Price Prospects for Major Primary Commodities,* vol. 4 (Washington, D.C.: World Bank, July 1982).

Notes: excludes Soviet bloc; includes China in developing countries; neg. = negligible; n.a. = not available and of iron and steel exports from both LDCs and Canada to industrial countries, have all expanded many times over.

The reason for the increasing shares produced in developing countries differs according to individual metals. The technology and the quality are similar in all countries, but processing in countries that do the mining saves transportation costs. Electricity is a major cost item for aluminum, and some developing countries, such as Brazil, have inexpensive hydroelectric power. Labor input costs less in most LDCs. Finally, meeting pollution abatement standards, which can be particularly costly in copper and zinc smelting, is usually a lower cost in developing countries. The relative costs for U.S. production of aluminum, copper, and steel are discussed in chapter 5.

Availability of Processing Capacities in the Late 1980s

Since the mid-1970s the decline in real prices of most nonfuel minerals has slowed the growth of processing capabilities in the United States, Canada, and Western Europe (in some cases it has meant the scrapping of old capacities without replacement). This is particularly true for aluminum, copper, lead, nickel, steel, zinc, and some of the ferroalloys. Analysts have suggested that a world industrial boom in the second half of the 1980s might lead to sharp increases in metal prices.[8]

Assessing the likelihood of such a development requires taking several different items into account. There is a large surplus processing capacity for most minerals in developed countries, but less of it in LDCs. As demand rose with the growth in industrial output, most real prices of minerals would again equal their full real economic costs.[9] This would probably happen well before normal capacity was achieved, and some additions to processing capacity could be made relatively quickly. Moreover, mine capacity is unlikely to present a constraint for most metals in the short term. Although speculative demand could sharply increase particular metals prices for a year or two, there is a reasonable chance that adjustments in processing capacities would be made without increases in real prices beyond historic levels. Indeed, World Bank projections show an actual decline in the combined index of real prices of metals and minerals between 1980 and 1995 (see table 2-5).[10]

To meet the demand for nonfuel minerals in the 1990s and beyond requires a substantial increase in mining and processing capabilities. This means that the mining industries will need large investments in the late 1980s and early 1990s because their facilities require many years to complete. In addition, exploration must expand to locate new mineral reserves for future decades. Whether investment will be adequate depends upon a number of factors, including the availability of capital in the mineral-producing countries.

To decide whether a marked shortage of investment is likely, it is necessary to have some idea of the financial magnitudes involved. Estimates of the investment needed to meet projected consumption of minerals in future decades are highly sensitive to projected growth rates in demand for individual minerals. In recent years, these projections have been revised downward (even by the same investigators) nearly every year, as have estimates of required investment for the next decade or two.[11] Estimating investment needs requires not only a determination of the net additions to productive capacity called for, but also an analysis of the capacities that must be replaced or modernized, including alterations for meeting new pollution abatement standards. Replacement requirements may run from one-third to over 100 percent of the net additions to capacity, differing according to commodity, the stage of production, and the producing country involved.

Although projections are subject to over- or understatement, the results of a recent World Bank paper on investment requirements, together with a study made by the Economic Research Department of Chemical Bank of New York, provide the best currently available estimates.[12] Both studies determined investment requirements for the same group of eight nonfuel minerals. The World Bank paper calculated investment requirements for the 1981–95 period, while the Chemical Bank study covered the 1981–2000 period. The latter projected lower rates of growth for the consumption of minerals than did the former (see table 3-2).[13] According to the World Bank, the replacement of old and the building of new capacity combined will require additional capacity equal to more than 50 percent of that available in 1980 for all eight minerals except nickel and tin (see appendix tables A3-4 and A3-5). Appendix table A3-6 shows how much investment was necessary to produce one additional ton of metal capacity in 1981 for each of the eight metals. Table 3-3 gives estimated investment requirements for mining and processing in the industrial and developing countries for the 1981–95 period based on the capital costs given in appendix table A3-6. The estimated total investment requirement is $151 billion in 1981 dollars, or $10 billion per year, of which 50 percent ($5 billion per year) constitutes the investment requirement for developing countries. These estimates, however, are likely to be on the low side because certain important costs including energy, labor, and pollution abatement, will probably rise. Outlays for infrastructure, including highways, railroads, power, and mining communities, are likely to add 10 to 20 percent to the capital requirements of actual mining and processing operations. These additions would raise capital requirements in LDCs to between $5.5 billion and $6.0 billion annually.[14]

By way of comparison, the Chemical Bank study estimates investment requirements for the same eight metals between 1981 and 2000 at about $126 billion (in 1980 dollars), or $6.3 billion annually—of which only $2.5 billion

TABLE 3-2. Consumption Growth in Nonfuel Minerals in Industrial and Developing Countries, for Selected Periods, in Percentage per Annum

	World Bank			Chemical Bank				
Metal[a]	1961–80 (actual)	1970–80 (actual)	1980–95 (projected)	1981–85 (projected)	1986–90 (projected)	1991–95 (projected)	1996–2000 (projected)	1981–2000 (projected)
Copper	3.9	2.5	2.6	1.00	2.00	2.50	2.75	2.06
Tin	0.4	−0.8	0.2	0.50	0.75	0.50	0.25	0.50
Nickel	7.6	2.2	2.4	1.50	2.00	2.50	2.75	2.20
Lead	4.3	2.1	3.2	0.00	1.50	2.00	2.25	1.44
Zinc	4.5	1.3	3.1	0.75	2.00	2.00	2.00	1.69
Aluminum	9.9	4.3	3.9	1.50	4.00	4.50	4.50	3.63
Iron ore	5.4	0.8	2.5	0.00	0.50	1.50	1.50	1.00
Manganese ore[b]	8.6	1.8	3.0	1.00	1.00	1.75	1.75	1.38
Average[c]	6.4	2.4	2.9	—	—	—	—	—

Sources: World Bank, *The Outlook for Primary Commodities,* Staff Commodity Working Paper no. 9 (Washington, D.C.: World Bank, 1983), p. 87. Gerald Pollio, "The Outlook for Major Metals through the Year 2000: An Updated View," *Journal of Resource Management and Technology* 12, no. 2 (April 1983): 114.

Note: excludes Soviet bloc countries

[a]Refined metal includes secondary material unless otherwise noted.

[b]manganese content basis

[c]weighted by the value of consumption in industrial and developing countries in 1980

annually represents expenditures in developing countries. Since the World Bank figures are in 1981 dollars, the inflation-adjusted equivalents in the Chemical Bank study would be $6.9 billion and $2.7 billion, respectively.

The investment requirements for the developing countries estimated by the World Bank would not be the same each year, but would rise from an average of $4 billion per year for the 1981–85 period to $5 billion per year in 1986–90, and to $6 billion in the final period—all in 1981 dollars (see table 3-3). Approximately half the investment requirements of the developing countries would represent foreign exchange expenditures, while the remainder would constitute local currency outlays.[15] Assuming that the foreign exchange requirements were obtained from abroad, this would require capital imports for the mining sectors of the mineral-producing countries of $2 billion per year for the 1981–85 period, $2.5 billion in the 1986–90 period, and $3 billion in the 1991–95 period (all in 1981 dollars).[16] These amounts may be compared with external financing of about a billion dollars per year during the 1970s.[17]

Providing $5 or $6 billion annually for capital investment in mining and processing in developed countries should be no problem, once metal prices have risen from their 1982–84 levels and existing plants are operating at normal capacity. There is, of course, the question of whether the investments will be made soon enough to meet the growing demand. But developing

TABLE 3-3. Estimated Investment Requirements for Mining and Processing in Industrial and Developing Countries

	Billions of 1981 U.S. Dollars				Percentage of Total
	1981–85	1986–90	1991–95	1981–95	1981–95
Industrial countries					
Mining	4	7	8	19	34
Processing	11	22	24	57	60
Subtotal	15	29	32	76	50
Per annum	3	6	6	5	—
Developing countries					
Mining	7	13	17	37	66
Processing	12	11	15	38	40
Subtotal	19	24	32	75	50
Per annum	4	5	6	5	—
Industrial and developing countries					
Mining	11	20	25	56	37
Processing	23	33	39	95	63
Total	34	53	64	151	100
Per annum	7	11	13	10	—

Source: World Bank, *The Outlook for Primary Commodities,* Staff Commodity Working Paper no. 9 (Washington, D.C.: World Bank, 1983), p. 93.

countries may face more severe problems. External financing of from $2 to $3 billion per year for the mining sectors may prove unobtainable, so that needed mining and processing expansion fails to happen. Compared with net capital imports of over $100 billion in 1981, the amounts projected for external financing of metal-producing capacity are small. The financial requirements for nonfuel minerals are, however, heavily concentrated in a few countries—Chile, Peru, Zaire, and Zambia (copper); Bolivia, Indonesia, and Malaysia (tin); Brazil, Guinea, and Jamaica (bauxite-alumina); Brazil, India, and Venezuela (iron ore); and Mexico and Peru (lead, zinc). Most of these countries now have severe external debt problems and there is some question about whether they will have much net external borrowing capacity for any purpose over the next five years. Moreover, capital imports for the mineral sector may be given a lower priority than capital imports for other purposes.

External financing for the mineral sectors in developing countries is most problematic when government-owned enterprises are involved. Investment in the private sector, mainly by large multinational mining firms, can usually be financed either directly by the parent firms or by borrowing under the sponsorship of multinational firms. It is likely, therefore, that the expansion of mining and processing capacities in developing countries will depend heavily on the willingness and ability of multinational mining companies to invest.

Multinational Mining Firms in a Changing Investment Climate

Throughout much of this century the world's mineral industries were dominated by multinational mining firms, whether American, British, Canadian, Belgian, French, or South African. The parent firms and their affiliates explored and mined in developed countries well before their exploration and introduction of large-scale mining in the LDCs. Most of the large mining complexes in the Third World today were established by these multinationals and they discovered and explored most of the large nonfuel mineral deposits, whether developed or not. These firms have operated almost exclusively in the nonfuel minerals industries, although with some minerals (such as aluminum) they are involved in fabrication as well.

Although the multinationals are still very important in world mining, three developments during the past two decades have modified their role. The first was the nationalization of foreign mining operations in the mineral-producing LDCs and the growth of state mining enterprises during the 1960s and 1970s. The second was the deterioration of the foreign investment climate in minerals in the developing countries. The third development was the merger of leading multinational mining firms with large petroleum and other firms, with a subsequent shift of management control away from those primarily experienced in nonfuel mining. These mergers occurred with increasing frequency in the late 1970s and early 1980s.

All three developments have had important effects on the flow of investment in exploration and development. The shift in mining industry ownership and management in LDCs to state enterprises has reduced both available investment and technology. The slowdown in foreign direct mining investment in the LDCs in favor of investment in developed countries such as Australia and Canada has also retarded Third World mineral industry progress. The loss of independence of many of the older multinational mining firms that had been taken over by giant petroleum and other firms whose principal business is not mining has affected managerial decisions and the flow of capital in ways in which net effect on investment in nonfuel minerals is yet to be determined.

The rest of this chapter reviews these developments to assess their impact on exploration and construction of mining capacity in both developed and developing countries. Will these developments retard or promote the growth of exploration, mining, and processing in the LDCs and how will they affect the long-run availability of nonfuel minerals?

In dealing with these and other questions relating to the change in ownership and control patterns in world mining, generalizations cannot cover all major nonfuel minerals. Moreover, changing ownership and control patterns will not necessarily have the same results for all developing countries. The economic, political, and social structure of African countries such as Zaire and Zambia differs from that of Brazil and Chile far more than the structure of the latter countries differs from developed countries such as Australia and South Africa. Thus, this assessment will take into account the effects of changing ownership and control patterns on both individual minerals and on countries at different stages of economic and political development.

The Decline in Foreign Direct Investment in Nonfuel Mineral Industries

The era of nationalization of foreign mining enterprises during the 1960s and 1970s has been well documented.[18] This period of extreme nationalism by host governments on the one hand and rigid adherence to traditional property rights by the multinationals on the other was followed by one of accommodation in a number of Third World countries during the late 1970s and early 1980s. Rather than expropriate the remaining foreign investments, most Third World governments have increased *control* over foreign-owned enterprises and recent mining legislation and contract terms for new foreign investments have reflected this trend. Although most LDCs no longer oppose foreign investment in their mineral resources, and, indeed, welcome investment on their own conditions, they have relied increasingly on state-owned enterprises for the development of their mineral resources.

Investment by multinational corporations in the mining and smelting industry of developing countries increased fairly rapidly after World War II but prior to the wave of expropriations beginning in the late 1960s. However, since 1970 both the book value of these investments and real annual capital expenditures by multinational firms in developing countries have declined sharply. Global data on multinational investments in developing countries by industry groups is scant and unreliable, but since U.S. multinational corporations account for well over half of this investment, U.S. data can serve as a proxy for all multinational corporations. The book value of U.S. direct investment in mining and smelting in developing countries in current dollars rose from about $750 million in 1950 to nearly $2.5 billion in 1968, more than a doubling of the book value in 1950 dollars.[19] In 1981 this investment remained about the same as in 1970, but after accounting for price rises, its real value in 1981 was less than half the 1970 level. Much of this decline in real value can be accounted for by expropriations in Latin America and Africa.[20] Foreign investment in mining by other developed countries experienced a similar decline during the 1970s.[21]

The book value of U.S. direct foreign investment in mining and smelting in developed countries rose from $2.7 billion at the end of 1970 to $4.9 billion at the end of 1981, but after taking account of the rise in prices, the real value declined by about one-third.[22] Thus, in terms of book value, U.S. direct investment in the mining industry of other developed countries fared considerably better than in developing countries.

Another measure of U.S. direct foreign investment in the mining industry is annual capital expenditure in this sector by majority-owned foreign affiliates of U.S. companies. Table 3-4 compares capital expenditure of this sort between developed and developing country recipients over the period from 1974 to 1982. Here, too, the relative decline was somewhat less for developed than for developing countries.[23]

There is also evidence that the portion of *total* investment from all sources (domestic and foreign) in minerals represented by investment in developing countries has been declining since 1970. Thus, on the basis of

TABLE 3-4. Capital Expenditures in the Mining and Smelting Industry by Majority-Owned Foreign Affiliates of U.S. Companies, in Millions of 1982 Dollars

	1974	1975	1976	1977	1978	1979	1980	1981	1982
Developed countries	1,450	1,422	1,089	650	506	677	1,080	1,009	530
Developing countries	599	545	369	145	126	182	356	333	159

Source: Survey of Current Business (Department of Commerce, Washington, D.C.), March and September 1976–83. Dollar investments deflated by the U.S. producer price index.

incomplete data, Philip Crowson estimates that over 20 percent of the total investment in minerals in 1970 was in developing countries, but that by 1980 the percentage had declined to about 10 percent.[24] This finding is broadly supported by the data on U.S. foreign investment in mining shown in table 3-4.

The Current Status of Multinationals in the Industry

Ownership and control of the world's nonfuel mineral production by multinational mining companies reached a zenith in the early 1960s, but in the past two decades their role has eroded. (Even in developed countries such as Australia and Canada, private domestic ownership and control have expanded relative to that of multinational firms based in the United States and Britain, and some governmental action favoring domestic ownership has fostered this trend.) Of major concern, however, is not simply the declining proportion of LDC mining capacity under multinational control, but the fact that in recent years they have not been increasing their investments in line with the long-term growth of world mineral demand.

Appendix table A3-7 lists the major multinational firms producing nonfuel minerals, the minerals involved, the location of their principal investments by multinationals, and their total assets. The companies listed in appendix table A3-7 account for well over half the non-Communist world production of gold, bauxite-alumina-aluminum, nickel, molybdenum, diamonds, lead, and zinc; over 40 percent of copper production; and a substantial portion of the world's output of iron ore, manganese, tungsten, chromium, vanadium, platinum-palladium, and tin. Although the companies' investments are found in a number of developing countries, the largest share of their mining assets is in developed countries.

The concentration of non-Communist world production of several important nonfuel minerals under leading multinationals is shown in table 3-5. Half or more of the bauxite-alumina-aluminum, refined nickel, tin metal, and refined lead and zinc, and over one-third of refined copper are controlled by multinational firms. Although not shown in table 3-5, U.S. steel companies control the vast bulk of their own iron ore supply; Western Europe is more than 80 percent self-sufficient in iron ore, much of which is also controlled by the steel companies. These data indicate that despite past nationalizations, multinationals still control a large share of non-Communist world production of major nonfuel minerals. This is important because multinational mining companies are in the best position to provide the capital, technology, and managerial skills to expand productive capacities in line with the growing demand for nonfuel minerals. Also significant, however, is the multinationals' diminished role in the LDCs, which contain a large share of

TABLE 3-5. Shares of Leading Multinational Companies in World Mineral Producing Capacity for Selected Minerals, 1977, in Percentage

Company	Mineral		
	Bauxite/Alumina/Aluminum		
	Bauxite Mining	Alumina Refining	Aluminum Smelting
Alcoa	20.8	22.4	13.0
Alcan	6.8	9.9	13.1
Kaiser	13.2	9.6	7.8
Rio Tinto-Zinc	6.6	2.0	2.1
Reynolds	6.0	9.6	9.4
Alusuisse	4.9	5.3	5.2
Pechiney	4.9	8.6	6.8
Noranda	1.3	<1.0	<1.0
Anaconda	<1.0	<1.0	2.4
Mitsubishi	<1.0	<1.0	2.7
Total	64.5	67.4	62.5
	Copper		
	Mining	Smelting	Refining
ASARCO	7.3	9.9	7.8
Kennecott	5.5	5.5	5.9
Rio Tinto-Zinc	5.5	2.8	2.6
Phelps Dodge	3.9	5.1	5.4
Anaconda	2.9	3.5	2.6
Newmont	2.7	2.7	<1.0
INCO	2.2	<2.0	<2.0
Mitsubishi	<2.0	3.6	3.0
Noranda	<2.0	2.6	4.9
AMAX	<2.0	2.5	2.8
Total	30.0	38.2	35.0
	Nickel (Refined or Ferronickel)		
INCO		36.8	
Falconbridge		15.3	
Le Nickel–SLN (French)		6.6	
AMAX		5.0	
Freeport		3.4	
Newmont (Sherritt-Gordon)		2.5	
Total		69.6	
	Tin Smelting		
Patino NV		29.6	
Billiton		10.7	
Rio Tinto-Zinc		8.5	
Union Miniere		2.1	
Total		50.9	

Continued on next page

TABLE 3-5—*Continued*

Company	Mineral	
	Zinc	
	Mining	Smelting
ASARCO	6.8	4.5
Noranda	6.1	3.6
Texasgulf	5.3	<2.0
Cominco	4.4	4.4
RTZ/CRA	3.8	4.7
AMAX	3.2	<2.0
Union Miniere	3.0	12.2
St. Joe Minerals	2.8	4.3
Cyprus (Amoco)	2.6	<2.0
Mitsui	2.3	5.5
Penarroya (French)	<2.0	3.8
Metallgesellschaft	<2.0	3.7
Anglo-American	<2.0	3.1
Total	46.3	49.8
	Primary Lead Refining	
ASARCO	7.7	
Penarroya	7.0	
Broken Hill	6.9	
AMAX	6.1	
St. Joe Minerals	6.0	
Cominco	4.5	
RTZ	4.4	
Union Miniere	3.7	
Preussag	3.5	
Total	49.8	

Source: United Nations, *Mineral Processing in Developing Countries* (New York: United Nations, 1980). Excludes centrally planned economies.

the world's unexplored and undeveloped mineral resources. For reasons given later in this chapter, the outlook for a major expansion of investment by the multinationals in these countries is not promising.

Effects of Recent Mergers on Investment in Nonfuel Minerals

Large U.S. multinational mining firms taken over by other (nonmining) firms during the late 1970s and early 1980s include Anaconda (by ARCO); Freeport Minerals (merged with McMoran); Kennecott Copper (by Standard Oil of Ohio); and St. Joe Minerals (by Fluor Corporation). (The large aluminum and steel companies are in a somewhat different category since their major objective in the production of ores and metals is to supply raw materials for their manufacturing operations.) Some of the large U.S. petroleum companies, such as EXXON, Standard Oil of Indiana, and Getty Oil (recently taken over by Texaco), have also organized mining subsidiaries operating here and abroad. These developments, which may well continue over the next few years, have made the traditional independent mining firms less important.

The mergers and the direct entry of petroleum firms into mining have provided certain advantages from the standpoint of future nonfuel minerals investment. Most important, the new partners in the mining business, such as petroleum companies, have large amounts of capital to invest. Due to low profits during the past decade, and to high capital costs for new mine and processing facilities, traditional mining firms had less equity capital for new mining ventures and relied heavily on borrowing. On the other hand, large conglomerates allocate their investments according to their perceptions of the most profitable opportunities among a range of investments. They are not as motivated as traditional mining companies by a desire to increase or maintain the firm's share of world mining. The traditional mining firm's staff includes a highly specialized group of geologists and mine managers whose policies aim to utilize these unique resources by finding and developing new ore bodies throughout the world. This was true even when mining was less profitable than other industries. The new mining subsidiaries must now compete for capital with parent company interests in other fields such as petroleum or manufactures.

One reason companies entering the nonfuel minerals industry have taken over extant companies rather than organizing their own mining subsidiaries is that traditonal mining companies have large undeveloped reserves. It is often cheaper and less risky to purchase reserves than to acquire them through exploration. Some traditional mining companies have also taken over their competitors, mainly to acquire reserves. It seems likely that mergers for the purpose of acquiring reserves will tend to reduce exploration. Fewer new entrants will explore to find reserves; fewer independent firms will explore to expand them.

The recent low prices and profits in mining and the poor outlook for the immediate future have disappointed the large companies that diversified their assets either by taking over existing mining companies or by forming their own subsidiaries. Investment in exploration and mining has declined substantially, particularly in the developing countries. Just how much of this decline can be attributed to the change in ownership structure brought about by takeovers is difficult to determine because other factors may have also reduced the attractiveness of such investment. Perhaps the answer will emerge when mineral prices rise and mining again becomes relatively profitable. It can then be determined whether the new managers allocate sufficient capital to meet growing requirements for minerals.

Controls over Foreign Investment in Developed Countries

The two largest developed-country mineral producers (excluding the United States)—Australia and Canada—exercise controls over new mining ventures by foreign investors, including takeovers. Both countries eventually want to reduce foreign control to less than 50 percent. Both subject new foreign investment projects to screening, but in Australia this applies to all new resource projects proposed by foreign controlled firms. These firms may undertake mineral exploration and subsequent development provided they propose an acceptable plan for eventual 50 percent Australian ownership and control. In Canada only new foreign enterprises, including takeovers, are screened; existing foreign concerns do not require special approval for new projects in the same industry. Foreign investment controls have not constituted a serious impediment to foreign investment in either country. In both countries substantial domestic capital is available to buy into mineral ventures initiated by foreign investors, and government controls have generally been administered so as not to deter new investment in exploration and development. In recent years both countries have approved the vast bulk of foreign investment proposals in their mineral sectors.[25] Moreover, the present or future majority private domestic ownership and control is less likely to interfere with the objectives of foreign investors here than in LDCs where domestic control is likely to be exercised by the government. Finally, large and efficient Australian, Canadian, and South African mining firms independent of foreign-based multinationals are quite capable of exploring and developing the resources of their own countries—and have also created successful mining ventures in a number of developing countries (see app. table A3-7).

The Outlook for Foreign Direct Investment in Third World Nonfuel Minerals

As has already been noted, future exploration and development of the rich mineral resources of the LDCs will depend heavily on the inward flow of external capital and technical and managerial skills. Even in countries such as Chile and Peru where there is a competent domestic mining industry, foreign direct investment will be needed to supply the large financial resources required.

The investment climate in a foreign country as perceived by potential foreign investors has many dimensions. It includes the tax regime, regulations relating to disposal of foreign exchange earnings by the investor, requirements for government equity participation, and government controls over a range of foreign company activities, including employment and training, domestic procurement of materials and equipment, environmental protection, and the provision of services and facilities to employees. In addition to conditions established by the government in legislation or mining contracts, investment climate includes the strength and militancy of labor unions, and the political and economic stability of the country. Chile has relatively favorable tax laws and other legal conditions affecting foreign investors, but economic and political conditions are shaky. Indonesia is relatively stable politically and economically, but its terms for investment in the mining industry have been generally unacceptable during the past decade. Peru has both high corporate taxes (up to 70 percent) and political instability, so there is little interest in investing there despite its rich mineral resources.

Widespread political and social unrest may portend the advent of a government that will change the fiscal regime affecting foreign investors and repudiate mining contracts negotiated by its predecessor. Economic instability may generate financial crises that result in restrictions on the transfer of profits to the foreign owners or limitations on spending foreign exchange for imports required by the mining enterprise. Labor unions may have a history of political strikes or of frequently violating union contracts with employers.

Unfavorable investment climate does not constitute an absolute barrier to foreign investment, but it does affect both the expected net return on capital and the probability of attaining it. If the expected before-tax return is very high, even an 85 or 90 percent tax on net profits—as with petroleum investments in Indonesia and Malaysia—may not deter investments.[26] If the expected net return from an investment is quite high, the investor may also be willing to accept relatively high risks, such as those represented by political instability. For instance, U.S. petroleum companies have made large investments in Angola, a country with a leftist government plagued by internal revolution and a border conflict with South Africa. Unlike petroleum investments in countries with large resources, exceptionally high returns in mining are rare. Also, the gestation period for large projects is longer than that for most involving petroleum, and it takes longer to repatriate capital after production begins. Therefore, mining investments in countries beset by economic and political instability or with high corporate tax rates are unlikely to prove attractive. Of the approximately twenty-five major nonfuel mineral exporting countries in the Third World, at least half have investment climates that most multinational mining firms would regard as unsatisfactory.

Trends in Third World Mining Contracts

Even where the general investment climate is relatively favorable, the terms of mining contracts offered by host governments are frequently unacceptable to foreign investors. Most mining contracts negotiated during the 1970s provide that ownership of natural resources remains with the state, and in many cases even the physical facilities at the mines will revert to the state when the long-term contracts (usually thirty years with the possibility of renewal) end. Agreements provide that investors may carry on exploration and mining activity under certain conditions and subject to certain government controls, and that they share revenues from production and sale of minerals with the state. Some current mining contract features that discourage foreign investment are discussed in the paragraphs that follow.[27]

An important factor in any investment is the prevailing fiscal regime, which involves royalties, rates of taxation on net profits, allowable depreciation, excess profits taxes, dividend-withholding taxes, and profit sharing due to host government equity participation. Although a contract's provisions reflect the host country's legislation, special tax treatment is frequently included, such as lower tax rates in the early years of operation, or accelerated depreciation allowances (all or a portion of capital outlays may be deducted from gross income in calculating taxable income).[28] There may also be guarantees protecting investors against any change in the fiscal regime applying to them during the life of the contract or for a certain number of years.

Prospective investors take into account all elements of the fiscal package in determining the risk-adjusted expected rate of return.[29] Thus the impact of a relatively high corporate tax rate may be moderated by accelerated depreciation or by the absence of a royalty paid on mine output or sales. The form in which taxes are levied is important to U.S. investors, since foreign taxes on net income count against domestic tax liabilities, whereas other kinds of taxes (e.g., royalty and export taxes) do not. A U.S. investor will generally not object to a tax regime similar in structure and rates to that at home. Fortunately, in most Third World countries normal corporate tax rates do not greatly exceed those of the United States.

Host governments frequently levy some form of excess-profits tax to limit after-tax profits in any one year. Foreign investors oppose such practices for two reasons. First, profits in the mining industry are subject to wide annual fluctuations, so that high profits in some years are necessary to balance low profits (or losses) in others. Second, multinational firms expect to balance low profits or losses in some countries with high profits in others, and they cannot be sure in advance which investments will perform well or poorly. Excess-profits taxes make their calculations all that more difficult.

A common feature in modern mining contracts is host government participation in the ownership of the company that operates the mine. Contracts frequently stipulate that the government has the option to acquire a certain percentage of the company's equity capital within a certain period after construction of the mine complex. Alternatively, the contract may grant a certain percentage of equity to the government without any payment on its part.

The 1976 Papua New Guinea (PNG) Ok Tedi agreement allowed the government to acquire 20 percent of the Ok Tedi mining company's equity. In 1982, following a feasibility study and final approval of the mining plan, but before the actual construction began, the government took this option. Equity payment was based on capital expenditures by foreign shareholders up to that time: this meant that PNG shared the capital risk after the feasibility study.[30] By and large, the Ok Tedi agreement was more attractive than mining contracts negotiated in Indonesia, Botswana, and other LDCs.[31]

Most multinationals resent mandatory equity participation for two reasons. First, it dilutes the foreign investor's equity and reduces the project's rate of return. Foreign investors who have to sell part of their equity after risky investment over several years find that the government's share is bought well below actual value.[32] The second reason is that some foreign investors regard LDC participation and their governments' presence as directors of a "foreign" firm incompatible with parent company control over its foreign subsidiaries. Some investors, however, find that government participation often provides it with a better understanding of the company's problems and gives the government a direct interest in its profitability. In a regional workshop on mining attended by representatives from both host governments and multinational companies in Asian and Pacific countries, several multinational representatives said that (minority) government equity participation was advantageous in the management of mining subsidiaries.[33] However, most companies are not willing to accept majority government ownership. In some cases this resulted in a multinational company's loss of control over the mining company. (In Zambia, for instance, after it acquired 51 percent ownership of American, British, and South African mining companies.)

LDC governments have recently insisted that the exploitation of their mineral resources make a contribution to economic development, especially in the area in which a mine is located. To achieve this and other objectives, they write contracts providing for considerable control over policies and operations. Modern contracts often require that detailed investment plans be submitted for government approval before mine and infrastructure construction begins. Such plans include: capacity of mine and processing facility, construction of infrastructure (roads, power and water, ports, housing and public service facilities for miners), disposal of waste, pollution control, and the schedule for completion.

LDC government approval of such plans usually requires months of negotiation with officials. Once the plans are approved there may be further negotiations about their implementation. Developing countries are increasingly concerned with the environmental impact of mining and smelting projects, and they often require stricter standards for foreign investors than for domestic investors. One of the most comprehensive environmental protection provisions is found in the PNG Ok Tedi contract. It deals with disposal of overburden and tailings, air pollution, the effects of mining operations on rivers and streams (including aquatic organisms) and its effects on farming, tribal life, and the general sociology of the region.

Recent mining agreements provide for maximum employment of host country nationals and for training nationals to assume positions at all levels. In some cases there are time schedules according to which specific percentages of nationals must be employed by job classification—up to and including managerial and professional positions.

Foreign investors can often agree with national employment and training programs because hiring nationals is usually cheaper than hiring foreigners with the same qualifications. On the other hand, difficulties arise when companies do not meet the time schedules, and employment of underqualified workers diminishes productivity. (Workers with adequate backgrounds for certain activities are frequently not available.) In addition, once they have been trained, local workers can find higher wages and better working conditions in other sectors of the economy, so that the training process never ends. Of course, this benefits the economy by raising levels of skill, but it often presents serious problems for the company. Governments frequently hamper operations by denying work permits to foreigners even when qualified nationals are unavailable.

Virtually all contracts obligate investors to buy host country equipment, materials, and services, importing only what cannot be purchased locally. Such provisions aim to foster domestic industry and agriculture. Most foreign subsidiaries find it useful to promote local supply facilities, but price, quantity, and delivery time may be inferior to what can be obtained abroad.

Multinational mining companies generally have accepted their host governments' perception of their relationship as a partnership geared to developing natural resources in accordance with national economic and social objectives. How provisions of a contract are enforced, however, may cause problems for both parties, especially if serious delays in production arise or if satisfaction of special political or economic interests (including those of labor unions) are paramount. The way a government deals with foreign investors often reflects its own efficiency and stability.

The risks of mining investment in politically volatile countries—notably in Latin America—are heightened by two legal principles held by many developing countries. One is that a contract with a private firm may be altered by the government whenever conditions change (or are alleged to have changed) from those existing when the contract was negotiated. This principle, referred to as *rebus sic stantibus,* is at odds with the legal principle of sanctity of contracts, *pacta sund servunda,* that underlies contract law in developed countries.

The other legal principle (adopted by virtually all Latin American states) is that contract disputes between the government and foreign investors must be settled by national courts, not by international arbitration. Adherence to this principle explains why only two Latin American nations (El Salvador and Paraguay) are members of the International Center for Settlement of Investment Disputes (ICSID) sponsored by the World Bank. (Most developed countries, and a number of Asian and African LDCs, belong to ICSID; submission of disputes to ICSID is explicitly provided for in some mine contracts.)

The legal principle of *rebus sic stantibus* often makes contracts little more than a frame of reference for continual renegotiation. As a consequence, foreign investors often give more weight to the reliability and stability of a government than to legal technicalities. When a country has a history of political instability, foreign investors face the likelihood of dealing with several quite different political regimes over the life of a twenty-five- or thirty-year contract. Moreover, economic instability threatens investors' ability to repatriate profits and make necessary purchases abroad.

Foreign investment in mining will continue to thrive in LDCs with a reasonable degree of political and economic stability. These countries include Brazil, Colombia, Indonesia, Malaysia, PNG, and Thailand. If a moderate democratic government were established in Chile, that mineral-rich country would probably enjoy a great deal of additional foreign investment. Countries like India, Mexico, and Venezuela have a long history of political stability, but they do not permit majority foreign ownership of mining enterprises. Foreign investors in mining do not like to accept a minority equity position. However, they sometimes do accept such a position, especially if they are assured of management control and if they have a veto regarding certain decisions that affect their interests. For example, Kaiser Aluminum has only a 49 percent equity interest in its bauxite operation in Jamaica, but has managerial control. In Brazil new foreign investment frequently takes the form of a joint venture with a state mining enterprise, a kind of venture that has proved to be relatively safe. China has recently shown interest in joint-venture contracts to develop its natural resources. For most other Third World countries, foreign direct investment in mining will probably be small, at least until they establish a record of stability.

Third World State Mining Enterprises

Three decades ago there were very few state-owned mineral enterprises in the LDCs. Almost all output was the province of multinational and private domestic mining companies. By 1980 much of the mineral output of developing countries was owned and controlled by government mining enterprises. Table

3-6 shows those LDCs with majority-owned state mining companies and the share these enterprises have in the country's total production of major minerals. Appendix table A3-8 lists the major Third World state mining companies and their productive capacities. In 1980 about 62 percent of the copper mining capacity in developing countries was held by government-dominated companies. Corresponding percentages were about 70 percent for iron ore, 33 percent for bauxite, and 40 percent for tin.[34]

In many LDC mining enterprises the government holds less than 50 percent equity. This minority participation entails varying degrees of government involvement in operations. In most of these joint ventures, multinational enterprises maintain operational control, while major policies are generally decided upon jointly. The degree of control does not necessarily correspond to

TABLE 3-6. LDCs in Which the State Holds 50 Percent or More of the Equity in Mining Enterprises, 1980

Mineral	Country	Approximate Percentage of Output Produced by State Enterprise
Bauxite	Ghana	55
	Guyana	100
	Jamaica	51
	Indonesia	100
	India	73
	Guinea	58
Copper (mining and smelting)	Brazil	100
	Chile	90
	India	65
	Malaysia	100
	Peru	20
	Zaire	100
	Zambia	60
Iron ore	Brazil	60
	Chile	100
	India	50
	Mauritania	100
	Peru	100
	Venezuela	100
Phosphate	Morocco	100
	Tunisia	100
Tin	Bolivia	100
	Indonesia	100
	Malaysia	25

Sources: Marian Radetzki, *State Mineral Enterprises* (Washington, D.C.: Resources for the Future, 1985); Walter C. Labys, "The Role of State Trading in Mineral Commodity Markets," in *State Trading in International Markets,* ed. M. Kostecki (New York: St. Martin's Press, 1982), chap. 4; and American Bureau of Metal Statistics, *Non-Ferrous Metal Data 1980* (New York: Port City Press, 1981).

the percentage of government equity ownership.[35] For example, Liberia's 35 percent ownership of the LAMCO iron ore joint venture does not mean the government exercises much control over it. Kaiser Aluminum and Reynolds Metals control their bauxite subsidiaries in Jamaica with only 49 percent of equity in each (see app. table A3-7).

About half the LDC state mining enterprises listed in appendix table A3-8 were originally developed and owned by multinational companies. These facilities still produce about two-thirds of the developing countries' copper and bauxite output and about one-fourth of their iron ore. An important exception is Brazil's Compania Vale do Rio Doce (CVRD), created by the government in 1942 and now the largest iron-ore producer in the world. Some of the more recent mining projects in which state enterprises have substantial equity, such as Brazil's Mineracao Rio do Norte (MRN, bauxite) and Colombia's Cerro Matoso (nickel), began as joint ventures with multinational mining firms.

Most steel companies in developing countries are government owned; few have been acquired by nationalization. Usually these enterprises serve only the domestic needs, but a few of them—notably in Brazil and Korea—are quite competitive with industrial countries in the world market. As of 1980, LDC exports of iron and steel represented only about 6 percent of world trade in these commodities, but that share, by all indications, will continue to grow.

Government policy in some developing countries aims at *domestication* of ownership and control rather than at state ownership and control as such. For example, the Mexican government has eliminated majority foreign ownership and control of large mining enterprises. Two of Mexico's largest copper mining companies were originally established by Anaconda and ASARCO. Anaconda was induced to reduce its equity in Minera de Cananea (currently Mexico's largest copper producer) to 49 percent, while ASARCO had to reduce its equity in Industria Minera Mexico (formerly ASARCO Mexicana) to 34 percent.

Increasing LDC ownership and control of their own nonfuel minerals industries raises several issues. One is whether these state enterprises have the financial and technical capacity needed to develop their mineral resources at a pace consistent with growth in world demand. A second concerns how state mining enterprises affect world markets. (This includes the tendency of state enterprises to produce at capacity levels even when world market prices are well below production costs on the one hand, and the chance they might form international cartels on the other.) A third concern is whether government might so monopolize mining that foreign enterprise cannot enter.[36] The final issue has to do with how international mining firms might integrate with state mining enterprises in the LDCs.

State Mining Enterprises: Technology and Financial Capacity

Because most of the world's large ore bodies in developing countries were first explored by multinational mining firms, which also developed most of the large operating mines, it is open to question whether state mining enterprises can explore and develop new ore bodies commensurate with their mineral resources. Exploration and development are essential if the growing world demand for minerals is to be met. Exploration is a high-cost, high-risk venture, and development of a large mine often costs billions of dollars. Exploration requires technology frequently not available to state mining concerns; development requires technology and financial resources, most of which must come from abroad.

Government-owned mining companies vary as to technical capacity, workers' experience, and managerial talent. Some of them inherited an exemplary managerial and technical staff when the mines were nationalized, since foreign-owned enterprises usually reduce their expatriate staff to a handful of people after several decades of operation.[37] It is not uncommon for the manager of a foreign-owned mine to be a national. For example, the manager of Southern Peru Copper Company's Cuajone mine, one of the largest in the world, is a Peruvian. When Zambia nationalized its copper industry, the managerial staff stayed on under contracts between the nationalized companies and the foreign companies that had formerly held majority interest. Even after these contracts were denounced by the Zambian government, most of the expatriate managerial and skilled personnel remained.

Countries lacking a well-developed mining industry would seem to need foreign investment for the exploration, creation, and operation of mines, and for the marketing of their products. However, this idea is not accepted by many who seek to eliminate foreign influence in the resource industries of developing countries.[38] They believe that LDCs can contract with foreign companies to construct "turnkey" projects and to operate them under limited-term management contracts that provide training for nationals at all levels.

All nonfinancial activities required for mining, from start to finish, can be subcontracted to foreign companies with appropriate technology and experience. Even multinational mining companies contract with exploration firms for geological studies; with large firms such as the Bechtel Corporation to construct mines; and with specialized companies for training programs. The same firms that undertake projects for international mining companies also work under contract for state mining enterprises. Nevertheless, certain important functions cannot be done as well by outside contractors lacking equity in the mining operation as by an experienced international mining company.

First, subcontractors do not risk capital. They are usually paid in advance for what they do or as the project proceeds. In any case, payment does not depend upon the success of the enterprise. International mining companies, on the other hand, risk capital at several stages: for exploration (perhaps involving hundreds of millions of dollars); for a feasibility study on a mining project (which may also cost millions of dollars); and for the equity to finance construction.

Second, even though a state mining enterprise can hire international firms to undertake exploration, feasibility studies, and construction, overall management remains in question. This requires experienced managers. Officials in an LDC's ministry of mines can seldom perform this essentially entrepreneurial function effectively, even though they may hire advisers to assist them.

The third function that subcontractors cannot handle is the mobilization of external capital involving perhaps billions of dollars. If a country has surplus revenues (as Saudi Arabia or Kuwait, for example), it can finance a large mining project itself. However, few developing countries can do this. Multinational mining subsidiaries usually have a high proportion of external debt: loans are for the most part from international commercial bank consortia that expect cash flow from exports to service the debt.[39] Commercial bank lenders usually require completion guarantees by the company establishing the subsidiary. Completion guarantees are in turn evaluated in terms of the multinational mining firm's proven record for project completion or—if necessary—restitution to creditors if the project should fail.

Unlike well-known internationals, state mining enterprises rarely have either the financial resources or the reputation needed to obtain large external loans. Moreover, external creditors find it difficult to enforce completion guarantees by a state enterprise. Reputable multinational mining firms, by contrast, have almost always honored such guarantees. Either they provided the additional capital needed to complete the project (when cost overruns were the problem), or they repaid creditors (when the project could not be completed).[40] Between 1981 and 1995, developing countries will require financing on the order of $5 billion per year (1981 dollars) to meet the projected demand for nonfuel minerals. Although data are not precise, it looks as though state mining enterprises would have to raise close to half that sum.[41]

The financial requirements of LDC mining enterprises compete with other economic sectors. Because debtor nations are unlikely to attract much foreign capital over the next decade or so, state mining enterprises will probably not gain more than a fraction of the estimated $2.5 billion per year needed to replace depleted ore bodies, to upgrade outmoded plants, and to expand production to meet projected increases in world consumption of nonfuel minerals. International development agencies such as the World Bank cannot supply more than a small fraction of the needed capital.

If LDC nonfuel mineral producers are to maintain their share of world export markets, they will need substantial foreign direct investment. Although foreign investors in LDC mining enterprises themselves rely on borrowed capital, they do so with the confidence of creditors about the performance and financial solvency of multinational companies. The total financing required will be small compared to the available international loanable funds.

State Mining Companies in World Markets

The behavior of state mining companies in marketing their products has been subject to two criticisms inconsistent with one another. Some argue that state enterprises do not reduce production and sales when slack demand and low prices should dictate such a course. As a result, prices decline even further. The second criticism is that they organize cartels to raise prices above the long-term equilibrium level. There is some validity to the first charge, but, for reasons discussed in chapter 7, there is little basis for the second.

The production and marketing practices of state mining enterprises have been cited as important reasons for the sharp decline of U.S. copper production in the early 1980s. Industry spokesmen argue that state mining companies can expand production even at a loss by reason of government subsidy.[42]

Although some state mining enterprises produce at a loss, others are quite profitable and pay the same taxes as private mining companies. For example, CODELCO reported net profits of $161 million in 1982, a year during which copper prices were low and most U.S. copper mining companies lost money. In Zambia, however, copper is evidently being produced at a loss.[43] Even when state enterprises produce at a loss, they are likely to maintain production because laying off workers costs almost as much as continuing to employ them.[44]

Contrary to popular opinion, mining enterprises in different countries are highly competitive with one another and pose little threat to consumers from cartel action. Only the International Bauxite Association (IBA) among the nonfuel mineral producers' associations exhibited sufficient cohesion for some of its members to follow common policies affecting export prices, but IBA's monopoly power is limited, and collusion among the major bauxite producers now appears remote. There are governmental producers' associations for other nonfuel minerals, but none has been able to exercise any monopoly power, and none is likely to do so in the foreseeable future.

The coordination necessary for cartel actions to raise price levels above true market levels is probably easier among multinational corporations than among state enterprises. For one thing, secrecy about collusion is hard to maintain. Agreements among state enterprises are more difficult to keep secret than those among international executives. The two most recent attempts to control international prices—the collusive action among mining firms to raise uranium prices and that of the DeBeers diamond cartel—were largely private in nature. However, multinationals will run afoul of antitrust regulations in the United States and other developed countries if collusive action becomes known. Governments may also take retaliatory trade action against governments that form cartels in raw materials.[45]

Has State Mining Reduced Foreign Investment?

Although most LDCs with large state mining enterprises permit foreign investment in their mining industries, foreign investors are often discriminated against in favor of the state enterprises, and in some cases cannot operate except in joint ventures with them. State mining enterprises often operate with great autonomy, and their officials tend to compete with private mining. This does not usually affect small domestic mining companies adversely because they have considerable political power of their own, but foreign mining companies have no domestic constituency. In most developing countries mining officials have great influence in the ministry of mines and can assure for the state enterprise the right to explore and develop the most promising mineral areas. In fact, governments frequently allot large areas of the country for the exclusive use of state mining enterprises.

A state mining company has great advantages over foreign competitors. It sometimes pays lower taxes, if any at all. It can also acquire government capital on better terms than a foreign private enterprise. State enterprises have been able to borrow from international financial agencies like the World Bank, or on the world capital market, backed by a government guarantee. Other advantages for state enterprises include their ability to develop land transportation and port facilities (or to have them provided by the government), to acquire imported materials and equipment more readily, and to get permits for personnel from abroad more rapidly. These advantages are often so great that foreign investors may not be willing to enter the country except in a joint venture with a state mining enterprise. For instance, foreign companies considering a mining investment in Brazil might be well advised to do so only as part of a joint venture with one of Brazil's state mining enterprises.

In Zaire the mining industry is dominated by the large state enterprise, Gecamines, which produces copper, cobalt, and other metals. Gecamines expanded with the aid of foreign loans, including substantial World Bank financing. Zaire welcomes foreign investment in mining; it is a country with very rich mineral resources. However, the only foreign mining firm operating there, Sodimiza, pulled out of Zaire in 1983 after losing over $200 million.

None of this proves that foreign investment in countries whose mining industry is state dominated cannot succeed. Multinational mining companies operate in Chile, Peru, and Indonesia along with large state mining enterprises.

Accommodation between State and Multinational Mining Companies

Joint ventures between state enterprises and multinationals provide foreign investors with an opportunity to transfer technology, management, and capital with relative security. Usually the partners' goals are similar. If state mining officials pursue legitimate business goals without interference from government administrators or politicians, serious conflict may not arise. Management is frequently in the hands of a committee made up of both state and foreign company representatives. The committee chooses the joint venture's manager. Majority equity ownership may be held by either party, but issues are almost always decided by consensus. Regardless of the distribution of equity, foreign investors usually have a veto over matters such as reinvestment of profits or major capital investment.

Some joint ventures were formed after partial nationalization of mining enterprises initially established by a multinational firm. This is the case with several of the state mining companies listed in appendix table A3-7. Joint venture firms in the truest sense of that term are those that began by combining capital and management of a multinational with a government agency or state enterprise. This is the case, for instance, with several Brazilian aluminum smelter and bauxite mining operations.[46] One of Brazil's large iron-ore mines is a joint venture formed by CVRD (51 percent) and eight Japanese steel companies who import the ore.[47]

Bauxite and iron ore are particularly good candidates for joint ventures because they are the raw materials required by the large integrated aluminum and steel companies that dominate world production and trade in these commodities. There are no organized markets for these materials (in contrast to copper) and the integrated aluminum and steel companies prefer to control output from ore through fabricated products. Joint ventures in LDCs provide integrated companies with a measure of control over supply and a market for the products made there.

Despite the advantages of joint ventures, the outlook for substantial increase in this form of foreign investment is not promising, except for a few countries like Brazil. State enterprises are often reluctant to share control over their mining and processing operations with foreign firms. One expert has pointed out that a state enterprise does not have the same freedom as a private company to negotiate joint ventures, because when the managers of state-owned enterprises have strong foreign linkages, "it helps them to slip loose from governmental control."[48] However, it is possible that some state enterprises may have to seek foreign joint venture partners to obtain capital because their governments have lost their ability to borrow abroad.

Conclusions

The adequacy of nonfuel minerals to supply global demand over the next decade will depend heavily on investment in new mine and processing facilities between 1985 and 1995. More than $10 billion (in 1981 dollars) will have to be invested annually if the project demand for nonfuel minerals in the market-economy countries is to be met. The developing countries will need half of this amount for their mineral industries to keep pace. Recent events suggest that exploration and mine development in Third World countries will not proceed as fast as their mineral resources warrant, measured against the projected growth in world demand. Multinational mining companies are not likely to be very eager to invest in these countries because of political and economic instability. Third World state mining enterprises, on the other hand, will have difficulty in obtaining the external capital needed to find new ore bodies and replace and expand their mine and processing capacities.

A relatively low level of investment in Third World mineral industries over the next decade will not necessarily mean a world shortage of nonfuel minerals or a substantial rise in their real prices between now and the end of this century. The developed countries have ample resources to substantially expand their production of all the most important minerals except tin. A rise in mineral prices to levels that made new investment in mineral industries profitable would mean increased capital flow to increase developed countries' capacity. However, the timing of investments in relation to growth in demand could result in short periods of high prices.

Appendix 3-1

Tables on Nonfuel Minerals: Uses, Production, Trade, and Investment

TABLE A3-1. Principal Industrial Uses of Major Minerals

Mineral	Principal Industrial Uses
Aluminum	Packaging (39%), transportation (20%), building (14%), electrical (8%), consumer durables (8%), other (11%)
Beryllium	Nuclear reactors and aerospace applications (40%), electrical (36%), electronic components (17%), other (7%)
Chromium	Metallurgical industry (52%), chemical industry (33%), refractory industry (15%)
Cobalt	Superalloys (37%), magnetic materials (16%), driers (11%), catalysts (10%), cutting and mining bits (7%), other (19%)
Columbium	Ferrocolumbium in the steel industry; alloys and metal in the aerospace industry
Copper	Fabrication as refined metal (80%), other (20%)
Diamonds (industrial)	Machinery (27%), stone and ceramic products (22%), abrasive (16%), construction (13%), mineral service (8%), transportation (6%), other (8%)
Gold	Jewelry and art (61%), industrial (29%), dental (9%), small bars (1%)
Lead	Batteries, gasoline additives (75%), construction, paint, ammunition (20%), other (5%)
Manganese	Ferromanganese in steel production; use in batteries, chemicals, and pig iron
Mercury	Electrical apparatus (52%), paint (14%), electrolytic production of chlorine and soda (21%), instruments (6%), other (7%)
Molybdenum	Iron and steel production (75%), machinery, oil and gas industry, transportation, chemicals, electrical (25%)
Nickel	Production of stainless and alloy steel (45%), nonferrous alloys (30%), electroplating (15%), other (10%)
Platinum group[a]	Automotive (33%), electrical (28%), chemical (15%), dental (9%), other (15%)
Silver	Photographic items (39%), electrical (29%), sterlingware, jewelry (14%), alloys and solders (7%), other (11%)
Steel	Construction (30%), transportation (24%), machinery (20%), oil and gas industry (9%), other (17%)
Tantalum	Electronic components (70%), machinery (22%), transportation (8%)
Tin	Containers (25%), electrical (17%), construction (13%), transportation (14%), other (32%)
Titanium[b]	Titanium metal used in jet engines and aerospace uses (60%), chemical processing industry (20%), production of steel and other alloys (20%)
Tungsten	Metalworking and construction machinery (72%), transportation (11%), lighting (8%), electrical (5%), other (4%)
Vanadium	Chief use as alloying agent for iron and steel; also important in production of titanium alloys and catalyst for production of sulfuric acid
Zinc	Construction materials (40%), transportation (20%), machinery (12%), electrical and chemical (15%), other (13%)

Source: Bureau of Mines, *Mineral Commodity Summaries 1983* (Washington, D.C.: U.S. Department of Interior, 1983).

[a]includes platinum, palladium, iridium, osmium, rhodium, and ruthenium

[b]produced from ilmenite and rutile

TABLE A3-2. Geographical Distribution of Mine Production and Reserve Base for Selected Minerals, 1980, in Percentage of Total World Production and of Reserve Base

	United States		Canada		Australia		South Africa		Other African Countries	
Mineral	Production	Reserves	Production	Reserves	Production	Reserves	Production	Reserves	Production	Reserves
Bauxite	2	<0.005	—	—	31	21	—	—	15	26
Chromium	—	—	—	—	—	—	35	68	6	30
Cobalt	—	10	5	1	5	2	—	—	67	52
Columbium	—	—	11	4	—	—	—	—	2	3
Copper	15	18	9	6	3	3	3	1	14	13
Diamonds (industrial)	—	—	—	—	—	16	17	8	49	63
Gold	3	1	4	3	—	—	56	58	—	—
Ilmenite	10	7	18	25	27	7	7	15	—	—
Iron ore	8	5	5	9	11	11	3	4	2	1
Lead	16	27	8	13	11	14	—	—	—	—
Manganese	—	—	—	—	7	6	22	41	8	3
Mercury	16	7	—	—	—	—	—	—	16 est.	8
Molybdenum	63	54	11	6	—	—	—	—	—	—
Nickel	2	5	25	14	—	—	—	—	—	—
Platinum group	<0.005	1	6	1	—	—	45	81	—	—
Rutile	n.a.	1	—	—	69	7	11 est.	4	15	2
Silver	9	22	10	19	—	—	—	—	—	—
Tantalum	—	—	23	8	14	21	—	—	10	22
Tin	n.a.	1	—	—	5	4	—	—	2	5
Tungsten	5	9	7	15	6	4	—	—	—	—
Vanadium	12	1	—	—	2	2	35	42	—	—
Zinc	6	20	16	26	9	10	—	—	—	—

Table A3-2—*Continued*

Mineral	Other Developing Countries		Other Non-Communist Countries[a]		Communist Countries		Percentage of World Production
	Production	Reserves	Production	Reserves	Production	Reserves	
Bauxite	28	30	16	20	8	2	Guinea (15%), Jamaica (14%)
Chromium	10	<0.005	12	1	37	1	Philippines (6%), Zimbabwe (6%)
Cobalt	5	9	5	12	13	13	Zaire (52%), Zambia (11%)
Columbium	86	93	1	—	n.a.	n.a.	Brazil (86%)
Copper	25	32	10	15	21	12	Chile (14%), USSR (12%)
Diamonds (industrial)	—	—	6	5	28	8	Zaire (32%), USSR (28%)
Gold	—	—	16	17	24	16	
Ilmenite	7	20	22	24	9	2	Norway (17%)
Iron ore	18	27	12	12	38	32	Brazil (12%), USSR (28%)
Lead	10	5	30	24	25	17	Peru (5%), Mexico (4%)
Manganese	15	1	4	1	44	47	USSR (39%), Brazil (8%), Gabon (8%)
Mercury	1	6	22 est.	57	45 est.	22	Algeria (16%), Spain (17%), USSR (32%)
Molybdenum	13	27	1	3	12	9	Chile (12%)
Nickel	11	25	34	37	28	19	New Caledonia (11%)
Platinum group	—	—	1	—	48	17	USSR (48%)
Rutile	2 est.	83	—	—	2 est.	2	Sierra Leone (12%)
Silver	28	20	29	16	24 est.	23	Mexico (14%), Peru (14%)
Tantalum	49	42	4	6	n.a.	n.a.	Brazil (29%), Thailand (19%)
Tin	67	58	1	3	21	25	Malaysia (25%), Thailand (14%), Indonesia (13%), USSR (15%)
Tungsten	18	7	16	7	48	58	China (28%), USSR (16%), Bolivia (6%)
Vanadium	—	—	10	3	40	52	USSR (28%), Finland (8%)
Zinc	13	4	33	30	23	10	Peru (8%), Mexico (4%)

Source: Bureau of Mines, *Mineral Commodity Summaries 1982* (Washington, D.C.: U.S. Department of Interior, 1982).
Notes: Data on total world mine production and reserve base are given in table 2-1; n.a. = not available.
[a]includes some developing countries

TABLE A3-3. World Trade in Nonfuel Minerals between Major Trading Countries and Regional Groups, in Billions of Dollars

		Destination						
Origin	Year	United States	Canada	Other Industrial Countries	Australia, New Zealand, South Africa	LDCs	Eastern Trading Bloc	World
United States								
Nonferrous	1973	—	0.48	1.14	0.04	0.44	0.02	2.12
	1980	—	0.70	3.44	0.06	0.97	0.03	5.20
Iron and steel	1973	—	0.74	0.58	0.06	1.18	0.04	2.60
	1980	—	0.68	0.42	0.06	1.89	0.06	3.11
Canada								
Nonferrous	1973	1.86	—	1.08	0.02	0.20	0.16	3.32
	1980	2.37	—	1.06	0.01	0.36	0.07	3.87
Iron and steel	1973	0.74	—	0.12	0.02	0.10	0.00	0.98
	1980	1.26	—	0.19	0.01	0.28	0.05	1.79
Other industrial								
Nonferrous	1973	1.28	0.20	11.28	0.12	1.24	0.90	15.02
	1980	1.91	0.08	20.44	0.13	2.47	1.10	26.13
Iron and steel	1973	4.58	0.54	23.66	0.86	8.96	5.38	43.98
	1980	5.47	0.51	27.92	0.78	15.78	7.04	57.50

Australia, New Zealand, South Africa								
Nonferrous	1973	0.24	0.00	1.04	—	0.28	0.00	1.56
	1980	0.88	0.01	1.53	—	0.90	0.00	3.32
Iron and steel	1973	0.12	0.02	0.30	—	0.34	0.06	0.84
	1980	0.44	0.04	0.44	—	0.90	0.15	1.97
LDCs								
Nonferrous	1973	1.32	0.04	6.44	0.00	1.04	0.22	9.06
	1980	2.42	0.05	5.91	0.05	1.75	0.40	10.58
Iron and steel	1973	0.56	0.02	0.44	0.02	0.84	0.06	1.94
	1980	0.92	0.05	1.16	0.05	2.30	0.20	4.68
Eastern trading bloc								
Nonferrous	1973	0.24	0.02	1.54	0.02	0.16	1.42	3.40
	1980	0.20	—	1.60	0.00	0.15	1.35	3.30
Iron and steel	1973	0.04	0.02	1.58	0.00	0.80	4.02	6.46
	1980	0.06	0.01	1.78	0.00	1.05	4.50	7.40
World								
Nonferrous	1973	5.94	0.74	22.54	0.20	3.36	2.72	34.40
	1980	7.78	0.84	33.98	0.25	6.60	2.95	52.04
Iron and steel	1973	6.04	1.34	26.68	0.96	12.22	9.56	56.80
	1980	8.15	1.29	31.91	0.90	22.20	12.00	76.45

Source: General Agreements on Tariffs and Trade, *International Trade 1977/78* and *International Trade 1982/83* (Geneva, Switzerland: GATT, 1978 and 1983), appendix tables.

TABLE A3-4. Production of Nonfuel Minerals in Industrial and Developing Countries, 1980 (actual) and 1995 (projected), in Thousand Tons[a]

Mineral[a]	1980			1995		
	Developing	Industrial	Total	Developing	Industrial	Total
Copper						
Ore/concentrates	3,888	2,262	6,148	5,910	3,610	9,520
Blister	3,147	2,774	5,921	5,500	4,000	9,500
Refined	1,550	6,110	7,661	2,780	8,560	11,340
Tin						
In concentrates	199	14	213	202	14	216
Metal	178	20	198	190	26	216
Nickel						
Metal	123	433	556	195	550	745
Lead						
Ore/concentrates	1,138	1,591	2,729	1,580	2,310	3,890
Metal	1,060	3,219	4,279	1,760	5,288	7,048
Zinc						
Ore/concentrates	1,714	2,963	4,677	3,310	4,070	7,380
Metal	1,047	3,410	4,457	2,514	4,846	7,360
Bauxite (in million tons)[b]	49	34	83	86	54	140
Alumina[b]	8,136	22,131	30,267	12,200	38,800	51,000
Aluminum	10,490	13,129	5,900	16,500	22,400	38,900
Iron ore						
Mining (in million tons)[b]	325	302	627	544	365	909
Pelletizing (in million tons)[b]	24	133	157	44	140	181
Manganese ore[c]	5,985	1,084	7,069	8,770	1,830	10,600
Ferromanganese[c]	1,414	2,688	4,102	2,865	3,615	6,480

Source: World Bank, *The Outlook for Primary Commodities,* Staff Commodity Working Paper no. 9 (Washington, D.C.: World Bank, 1983), p. 88.

[a]metal content unless otherwise noted

[b]gross weight

[c]manganese content

TABLE A3-5. Projected Capacity Replacement Requirements and Capacity Additions in the Nonfuel Mineral Sector in Developing and Industrial Countries, 1981–95, in Thousand Tons of Annual Production Capacity

	Developing			Industrial			Industrial and Developing		
Mineral[a]	Capacity to Be Replaced	Capacity to Be Added	Total	Capacity to Be Replaced	Capacity to Be Added	Total	Capacity to Be Replaced	Capacity to Be Added	Total
Copper									
Mining	535	2,010	2,545	525	915	1,440	1,060	2,925	3,985
Smelting	770	2,005	2,775	900	336	1,236	1,670	2,341	4,011
Refining	473	2,008	2,481	1,500	1,036	2,536	1,973	3,044	5,017
Tin									
Mining	94	12	106	5	1	10	103	13	116
Smelting/refining	16	0	16	0	0	0	16	0	16
Nickel									
Mining/processing	6	87	93	28	30	58	34	117	151
Lead									
Mining	503	471	974	572	800	1,372	1,075	1,271	2,346
Smelting/refining	217	777	994	568	2,405	2,973	785	3,182	3,967
Zinc									
Mining	486	1,710	2,196	998	1,093	2,091	1,484	2,803	4,287
Smelting/refining	80	1,519	1,599	653	1,261	1,914	733	2,780	3,513
Bauxite									
Mining[b]	3,939	39,740	43,679	737	24,420	25,157	4,676	64,160	68,836
Alumina									
Refining[b]	508	5,950	6,458	1,304	17,950	19,254	1,812	23,900	25,712
Aluminum									
Smelting	18	4,035	4,053	938	7,104	8,042	956	11,139	12,095
Iron ore									
Mining	62,700	190,000	252,700	55,000	55,000	110,000	117,700	245,000	362,700
Pelletizing	15,000	10,000	25,000	37,000	0	37,000	52,000	10,000	62,000
Manganese									
Mining	1,123	2,161	3,284	210	639	849	1,333	2,800	4,133
Processing	690	1,129	1,819	1,199	561	1,760	1,889	1,690	3,579

Source: World Bank, *The Outlook for Primary Commodities,* Staff Commodity Working Paper no. 9 (Washington, D.C.: World Bank, 1983), p. 90.

[a]tonnage in metal content unless otherwise noted

[b]tonnage in gross weight

TABLE A3-6. Estimated Investment Costs per Ton (metal content) of Annual Capacity in Industrial and Developing Countries, in 1981 Constant Dollars per Ton

	For Additional Capacity		For Maintaining Existing Level of Production	
	Developing	Industrial	Developing	Industrial
Copper				
Mining	6,000	7,000	2,700	3,700
Smelting	2,000	2,000	1,200	1,200
Refining	700	700	400	400
Tin				
Mining	4,500	5,000	2,200	2,400
Smelting	3,000	3,000	2,000	2,000
Nickel				
Mining/processing	37,000	36,000	26,000	25,000
Lead				
Mining	2,040	1,540	1,260	940
Smelting/refining	2,300	2,100	2,100	2,100
Zinc				
Mining	2,040	1,540	1,260	940
Smelting/refining	2,860	2,600	2,600	2,600
Bauxite				
Mining[a]	75	60	35	35
Alumina refining[a]	1,035	900	900	900
Aluminum smelting	3,500	2,900	2,900	2,900
Iron ore[a]				
Mining	63	53	25	21
Pelletizing	40	30	22	17
Manganese ore mining	158	160	60	62
Ferromanganese production	500	300	248	149

Source: World Bank, *The Outlook for Primary Commodities,* Staff Commodity Working Paper no. 9 (Washington, D.C.: World Bank, 1983), p. 92.

[a]The investment cost figures are estimated for a "gross" ton of annual capacity rather than for a ton of metal content. Metal content averages about 23%, 52%, 55%, and 70% for bauxite, alumina, iron ore, and iron pellets respectively.

TABLE A3-7. Major Multinational Firms Producing Nonfuel Minerals

Firm	Location of Principal Mining Investments	Principal Nonfuel Minerals	Total Assets (in billion U.S. $)
United States			
AMAX	U.S., Australia, Canada, Zambia, Botswana, South Africa, Philippines, Indonesia, PNG, U.K., Mexico, Dominican Republic	Molybdenum, copper, iron ore, lead, zinc, nickel, tungsten, silver, aluminum	5.1
Aluminum Co. of America (Alcoa)	U.S., Guinea, Australia, Jamaica, Dominican Republic, Brazil, Suriname, Mexico	Bauxite, alumina, aluminum	6.0
Anaconda (subsidiary of Atlantic Richfield)[a]	U.S., Mexico, Australia, Chile, Jamaica	Copper, silver, gold, zinc, aluminum, nickel, molybdenum	n.a.
ASARCO[a]	U.S., Mexico, Australia, Peru, Canada	Copper, silver, lead, zinc, gold, molybdenum	2.2
Amoco (subsidiary of Standard Oil Indiana)	U.S., Canada, Australia, PNG	Copper, molybdenum, other metals	n.a.
Exxon[a]	U.S., Canada, Spain, Chile, Australia	Copper, other base metals	62.3
Freeport/McMoran	U.S., Indonesia, Canada	Sulfur, gold, copper, nickel	1.7
Getty Oil[a]	U.S., Australia, Canada, Chile	Copper, other base metals	9.9
Hanna Mining	U.S., Canada, Brazil, Guatemala	Iron ore, nickel, bauxite, alumina, aluminum	0.5
Kaiser Aluminum and Chemical	U.S., Jamaica, Canada, Australia, New Caledonia, Ghana, Western Europe, Bahrain	Bauxite, alumina, aluminum, nickel	3.6
Kennecott (subsidiary of Standard Oil of Ohio)[a]	U.S., Canada, Australia, Mexico	Copper, molybdenum, gold, silver, lead, zinc, ilmenite	n.a.
Newmont Mining	U.S., Canada, Peru, South Africa, Chile, Indonesia, Philippines	Iron ore, cobalt, silver, gold, copper, nickel, lithium, molybdenum, vanadium	1.9
Phelps Dodge	U.S., Canada, Peru, South Africa	Copper, silver, gold, paladium	2.0
Reynolds Metals	U.S., Jamaica, Haiti, Canada, Brazil, Philippines, Ghana	Bauxite, alumina, aluminum, fluorspar	3.3
St. Joe Minerals (subsidiary of Fluor Corp.)[a]	U.S., Chile, Australia, Peru, Argentina	Lead, zinc, gold, copper, iron ore, silver	4.7
U.S. Steel Corporation	U.S., Canada, South Africa, Gabon	Iron ore, manganese, zinc	19.4
Australia			
Broken Hill Pty.	Australia, PNG, Indonesia	Iron ore, copper, tin, lead, zinc, gold, alumina, aluminum, manganese	6.9
CRA (subsidiary of Rio Tinto-Zinc, U.K.)	Australia, PNG, Malaysia	Copper, zinc, tin, gold, lead	4.5

Continued on next page

TABLE A3-7—*Continued*

Firm	Location of Principal Mining Investments	Principal Nonfuel Minerals	Total Assets (in billion U.S. $)
Belgium			
Union Miniere	Belgium, Canada, Brazil, U.S., Mexico, Spain	Copper, zinc, silver, gold, platinum/palladium	n.a.
Canada			
INCO	Canada, U.S., Indonesia, Guatemala, Australia, Brazil, New Caledonia, Mexico, Philippines	Nickel, copper, gold, silver, platinum/palladium, cobalt, magnetite	3.4
Cominco	Canada, Australia, U.S., Greenland, Japan, Philippines, Western Europe	Lead, zinc, silver, gold	2.7
Noranda Mines	Canada, U.S., Australia	Copper, gold, silver, lead, molybdenum, cobalt	4.6
Falconbridge Nickel	Canada, Norway, Dominican Republic	Nickel, copper, cobalt, iron ore, gold, silver	1.1
Aluminum Co. of Canada (Alcan)	Canada, Brazil, Jamaica, Guinea	Bauxite, alumina, aluminum	6.6
France			
Pechiney-Ugine Kuhlman (owned by French govt.)	France, Greece, Guinea, Australia, Spain, Netherlands, Canada, Cameroon	Bauxite, alumina, iron ore, aluminum	4.5
Le Nickel–SLN (subsidiary of Elf Aquitaine owned by French govt.)	France, New Caledonia, Cameroon	Nickel	n.a.
IMETAL	France and other Western European countries: Brazil, Morocco, Peru, Australia	Zinc, lead, silver, copper, other metals	1.7
Germany			
Metallgesellschaft AG	Germany, Australia, Canada	Lead, zinc, tin, tungsten	1.8
Preussag	Germany, Canada	Zinc, lead, copper, silver, mercury	1.3
Japan			
Mitsubishi Metal Corp.	Japan, Australia, Canada, Peru, U.S.	Copper, nickel, silver, gold, tin	1.5
Netherlands			
Billiton International Metals (subsidiary of Royal Dutch Petroleum)	Netherlands, Australia, Canada, Brazil, Suriname, Indonesia, Thailand, Peru, Western Europe, Colombia	Copper, nickel, tin, bauxite, tungsten, zinc, molybdenum	n.a.
Patino NV	Netherlands, Canada, Brazil, New Caledonia, Australia, Malaysia, Nigeria	Tin, nickel, cobalt, lithium	0.2
South Africa			
Anglo-American Corp.	South Africa, U.S., Canada, Botswana, Brazil, Zambia, Swaziland, Zimbabwe,	Copper, nickel, iron ore, platinum, manganese, tin, tungsten, diamonds, chro-	4.1

Continued on next page

TABLE A3-7—*Continued*

Firm	Location of Principal Mining Investments	Principal Nonfuel Minerals	Total Assets (in billion U.S. $)
	Western Europe, Australia	mium, silver, zinc, vanadium, gold	
DeBeers Consolidated Mines	South Africa, Botswana, Namibia, Mexico, Lesotho	Diamonds	4.8
Sweden			
Granges International	Sweden, Liberia, Canada, Saudi Arabia	Iron ore, phosphates	1.2
Switzerland			
Swiss Aluminum Ltd. (Alusuisse)	Switzerland, Australia, Canada, U.S., Sierra Leone, New Zealand, Guinea	Bauxite, alumina, aluminum, fluorspar, lead, zinc, copper, phosphates	5.0
United Kingdom			
Rio Tinto-Zinc	U.K., Australia, PNG, South Africa, Canada, Indonesia, Spain, Zimbabwe, Panama, New Zealand	Copper, lead, zinc, iron ore, alumina, aluminum, bauxite, silver	9.0
Consolidated Gold Fields	South Africa, Australia, U.S., Philippines, PNG, U.K.	Gold, copper, iron ore, platinum, silver, tin, ilmenite, titanium	2.1
Selection Trust (subsidiary of British Petroleum)[a]	South Africa, Australia, U.S., Sierra Leone	Copper, iron ore, gold, other metals	n.a.

Sources: Information on U.S. assets taken from *Moody's Industrial Manual* (New York: Moody's Investors Service, 1983), 1982 figures. Information on foreign company assets taken from *Moody's International Manual* (New York: Moody's Investors Service, 1982), 1981 figures, and "International 500," *Fortune,* August 22, 1983, 1982 figures. Information on location of principal mining investments taken from *Engineering and Mining Journal International Directory of Mining* (New York: McGraw-Hill, 1981), and company annual reports for location and minerals.

Note: n.a. = not available

[a]bulk of assets in petroleum or other industries outside nonfuel minerals

TABLE A3-8. Major Third World Mining Enterprises with 35 Percent or More of Equity Held by the Government, 1980

Country	Government Equity Share	Capacity (in thousand mt per year)	Other Equity Holders
		Bauxite (ore)	
Brazil			
Mineracao Rio do Norte	46%	3,400	Alcan (24%) plus minor foreign interests (30%)
Ghana			
Ghana Bauxite Co.[a]	55	300	British Aluminum Co. (45%)
Guinea			
Compagnie des Bauxites de Guinee	49	9,000	Halco Mining (51%)
Friguia	49	3,000	Pechiney, Noranda, British Aluminum, Alusuisse, Vereinigte Aluminum Werke (West German) (51%)
Offices de Bauxites de Kindia[a]	100	2,500	
Guyana			
Guyana Mining Enterprise[a]	100	2,500	
India			
Bharat Aluminum Co.	100	400	
Indian Aluminum Co.	45	500	Alcan (55%)
Indonesia			
PT Aneka Tambang	100	1,800	
Jamaica			
Kaiser Bauxite Co.[a]	51	4,200	Kaiser Aluminum and Chemical (49%)
Jamaica Reynolds Bauxite Partners[a]	51	3,100	Reynolds Metals (49%)
		Copper Mining (metal, blister, or refined)	
Brazil			
Brasileira do Cobre	100%	30	
Chile			
Corporacion Nacional del Cobre (Codelco)[a]	100	890	
Empresa Nacional de Minera (Enami)	100	25	
India			
Hindustan Copper	100	35	
Mexico			
Mexicana de Cobre	44	180	Private Mexican investors (56%)
Minera de Cananea[a]	52	65	Anaconda (ARCO) (48%)

Continued on next page

TABLE A3-8—*Continued*

Country	Government Equity Share	Capacity (in thousand mt per year)	Other Equity Holders
Peru			
Centromin[a]	100	34	
Mineroperu	100	33	
Zaire			
Gecamines[a]	100	662	
Zambia			
Zambia Consolidated Copper Mines[a]	60	704	British, American, and South African investors (40%)
		Iron (ore)	
Brazil			
Companhia Vale do Rio Doce (CVRD)	100%	57,000 (1983 prod.)	
Chile			
Compania de Acero del Pacifico[a]	100	6,935 (1978 prod.)	
India			
National Mineral Development Corp.	100	12,000	
Liberia			
Lamco[a]	37	10,600	Swedish and American investors (63%)
Mauritania			
Societe National Industrielle Miniere[a]	100	6,336 (1978 prod.)	
Peru			
Empresa Minera del Peru[a]	100	4,854 (1978 prod.)	
Venezuela			
Ferrominera Orinoco[a]	100	15,300 (1981 prod.)	
		Nickel (ore)	
Colombia			
Cerro Matosa	45%	700 (1984 est.)	Billiton (39%), Hanna (14%), Socal (1%)
		Phosphate (ore)	
Morocco			
Office Cherifien des Phosphates	100%	20,000	
Tunisia			
Compagnie des Phosphates de Gafsa	98	4,600	Public stock investors (2%)

Continued on next page

TABLE A3-8—*Continued*

Country	Government Equity Share	Capacity (in thousand mt per year)	Other Equity Holders
		Tin (content of ore)	
Bolivia			
Corporacion Minera de Bolivia (Comibol)[a]	100%	25	
Indonesia			
PN Tambang Timah	100	28	
Malaysia			
Malaysia Mining Corp.	71	15	Private investors (29%)

Sources: Marian Radetzki, *State Mineral Enterprises* (Washington, D.C.: Resources for the Future, 1985); Raymond Vernon and Brian Levy, "State-Owned Enterprises in the World Economy," in *Public Enterprise in Less-Developed Countries,* ed. Leroy P. Jones (Cambridge: Cambridge University Press, 1982), pp. 169–88.

[a]originally developed and wholly owned by a multinational mining company

4

Market Structure, Price Fluctuations, and Mineral-Producing Capacity

Most individuals react to changes in international mineral prices in terms of how their own welfare is affected: falling prices are welcome while rising prices are deplored. Thus the continuing decline in real prices of minerals over the past decade has curbed inflation and increased consumers' real income. But since most nonfuel mineral industries are competitive and over the longer run prices must cover full economic costs of production, price fluctuations not warranted by changes in real costs of production reduce the efficiency of production. The consequent increase in costs is ultimately borne by consumers.

Short-term cyclical fluctuations in prices for nonfuel minerals are to be expected in a competitive market. Although these fluctuations increase risks and therefore impair efficiency, they do not greatly affect production over time. On the other hand, a long period of declining real prices not accompanied by a decline in real production costs means low growth in capacity compared to growth in demand. What often happens is that investment, which is usually keyed to the historical rate of growth in demand for a mineral, may fall off sharply when growth does not meet that expectation. Oversupply of minerals thus entails an undersupply of the investment needed over the long haul. When existing productive facilities are in oversupply for two or three business cycles relative to effective demand, little or no additional investment is likely for exploration, new technology, or otherwise improved mining performance. A long period of low prices may also result in the abandonment of high-cost mines and the failure to maintain and replace plant facilities.[1] When prices and the outlook for future demand relative to productive capacities again become attractive to investors, it may require several years before new plants can be built, older facilities restored to production, and better technology purchased and installed. Meanwhile, prices may rise well above world economic costs until output catches up with demand.

A concrete example of this bust-and-boom phenomenon may be seen in the recent history of copper production. A reduction in the long-run growth in demand for it led to a decline in its real price that began in 1974 and continues in the 1980s. First there was the relative overcapacity generated by mines coming on stream after 1974 (initiated on the expectation that the earlier 4 percent annual rate of growth in demand would continue). Exacerbating this were the marketing practices of developing countries' new state enterprises. The consequent low level of investment in copper mining may result in a sharp rise in copper prices in the 1990s.

Both cyclical price volatility and longer periods during which prices of nonfuel minerals are much below or above full economic costs of production are consequences of competitive market forces. The demand for virtually all metals is highly inelastic in the short and medium term, since the cost of metal constitutes only a small portion of the value of the finished products. However, unless the sellers control the market to a significant degree, mineral supply is also highly inelastic. This is true because of the high proportion of overhead costs: sellers may continue to produce at prices below their total costs as long as their variable costs are covered. On the other hand, it may take several years to increase capacity enough to meet an increase in demand, during which time prices may be well above what they would have been were productive facilities adequate. Given the inelasticity of both demand and supply, if the market for a metal were purely competitive in the sense that no organized group greatly influenced its price, even small shifts in demand might result in very large changes in price.[2]

It is apparent, therefore, that how the market for a given mineral is structured has a great influence on both short-term price fluctuations and long-term real price trends. For example, if production is highly concentrated among a few firms that adjust their output to demand in an oligopolistic manner, the mineral's price would not fluctuate greatly and would usually not fall below full economic cost at normal capacity. ("Acting in an oligopolistic manner" here means that individual companies do not cut prices in order to maintain their market share at the expense of other firms.)[3] On the other hand, if there are a large number of firms in a purely competitive market, each selling as much as it can produce regardless of the market price, the latter will fluctuate widely in response to demand. Thus an essential condition for price stability is the willingness of firms to cut back production when demand goes down. Oligopolistic companies may also forego sharp price increases even when demand for their output exceeds their capacity to satisfy it. They may do this, for example, by rationing how much of their output they will supply to customers. Maintaining reasonable price stability for a product serves in part to discourage consumers from shifting to substitute materials. It also builds customer loyalty against times when producer prices are above those available on the open market. In the U.S. copper industry, for example, producer prices for almost every year between 1964 and 1974 were *below* London Metal Exchange quotations; they have been *above* the exchange price for most of the late 1970s and the 1980s.

It is something of a dilemma to choose between oligopolistically stable mineral prices and competitively volatile ones. On the one hand, prices that depart substantially from long-run equilibrium levels often result in overcapacity during some periods and undercapacity in others. This means that consumers sometimes obtain minerals at less, sometimes at more, than full costs.[4] On the other hand, competition is desirable, not only because it prevents consumers from being exploited by producers, but also because it promotes efficiency and technological advances that ultimately lower costs. There is no wholly satisfactory way of dealing with this dilemma. Government price controls and production planning not only have a poor record for promoting economic efficiency, but since we are dealing with *international* markets, price stabilization is, for reasons discussed in chapter 6, exceedingly difficult to achieve. In my opinion, any "solution" that impedes worldwide competition is likely to create more problems than it alleviates.

The Structure of Nonfuel Mineral Markets

The market structure for nonfuel minerals is complex. Markets differ for products at each stage of production—for ores and concentrates, and for separate categories of refined metals (including alloys such as ferromanganese). There are national or regional markets in major industrial countries, and there is an international market for most refined metals. There is no single competitive market for any mineral product encompassing all transactions, both national and international; even though all these markets influence one another, they are only partially integrated.

There are open markets for most refined metals but not for ores and concentrates sold to smelters or refineries, usually on a contract basis. Aluminum, copper, gold, lead, silver, tin, and zinc are traded on one or more commodity exchanges, in either "spot" or "futures" transactions. The principal metal exchanges are the London Metal Exchange (LME) and the New York Commodities Exchange (COMEX). Such markets reflect world demand and supply although their transactions account for only a small percentage of total volume.[5]

The metal exchanges are nevertheless important in the determination of world prices for the commodities traded. First, market quotations there serve as reference prices for sales contracts between producers and consumers or processors. Second, futures markets operated by commodity exchanges allow consumers, producers, and dealers to hedge against price fluctuations. For instance, producers of fabricated copper products may want to be sure of how much they must pay for refined copper before negotiating a contract to deliver their output at a set price. If so, they can negotiate a futures contract for the amount of refined copper needed. Producers or dealers supplying copper can also be assured of the price they will receive by negotiating a futures contract.

Futures transactions account for a much higher volume than spot transactions on commodity exchanges.

Most domestic and international trade in metals takes place under annual or multiyear contracts for the delivery of specified amounts (or maximum and minimum amounts) of the metal by producers. The actual price paid on delivery may be the price set by the producer (the producer price) or a price geared to the current spot price on one of the exchanges. Producer prices change less frequently than commodity exchange prices; for some metals the former may long remain significantly higher or lower than the latter. Producer prices set by large mining firms in the United States and Canada tend to move together because of "price leadership," which means that firms follow the leadership of the firm that initiates a change in the price, rather than trying to undercut the leader by selling more cheaply. Producers can charge significantly higher prices than those on the commodity exchanges because of a combination of oligopolistic control by a few large producers and the advantages their customers see in having relatively stable prices over an extended period. (Consumers continue to buy under an annual quantity contract at a producer price that may be higher than the exchange market price because this assures them of that quantity of the metal at a relatively stable price compared to the variations likely on the commodity exchange.) There are also quality differences even in metals with the same standard specifications, and buyers want to be sure of a continuous supply of the same quality metal identified with a particular producer. In recent years, however, producer prices for some metals have changed more frequently and bear a closer relationship to commodity exchange prices. This has increased price volatility somewhat.

Besides these two kinds of arrangements for minerals trading there is also a *merchant market*. It consists of middlemen who buy metals such as aluminum, copper, and zinc, frequently from secondary refineries (i.e., refineries producing from scrap materials) or from small primary producers, and sell to consumers without long-term purchase contracts or who want to acquire a portion of their supplies outside these contracts. Prices on the merchant markets may differ substantially from producer prices. For metals traded on commodity exchanges, merchant prices tend to follow prices quoted there. The merchant market becomes particularly active when producers are unable to fill all the orders placed with them, and also when a mineral is in relative oversupply and open market prices are thus considerably lower than producer prices. For metals not quoted on commodity exchanges, only producer price quotations may be available. Thus a large number of transactions may take place at unknown prices since there is no regular system to record merchant market transactions.

Some large firms (notably those in aluminum and steel) are integrated from mining through fabrication. In such cases the product at various stages of production is simply transferred from one part of the firm, or from one controlled subsidiary, to another; only the finished product is sold.

The marketing arrangements for individual nonfuel minerals do not fall into airtight categories since the same commodity may be sold in different types of markets. Appendix 4-1 gives a brief summary of these arrangements.

Market Structure and Price Volatility

The price volatility of a nonfuel mineral depends largely on the kind of market on which it is traded. As already noted, the more competitive the market, the greater the cyclical price volatility and the more likely prices are to depart from the long-run equilibrium level for a long time. Concentration in the production of a metal facilitates the maintenance of more or less uniform producer prices. When demand falls prices may not fall correspondingly, since producers can either reduce output or add to inventory rather than lower prices enough to sell all they can produce. When demand surges prices may remain relatively stable because producers can draw upon their inventories or, in some cases, ration output.

There are international producer prices as well as national or regional ones. Uniform producer prices tend to prevail when only a few countries account for most of a mineral's world supply. For example, the marketing agent for cobalt produced by Zaire's state enterprise establishes a cobalt price that is more or less followed by Zambia and other countries producing that metal. Likewise, South African chromium producers set prices for chromite ore and ferrochromium that other producers tend to follow.

Metal traded on a commodity exchange such as the LME or COMEX tends to be volatile in price. However, only a small portion of all metals transactions take place there; most are based on contracts. There exists, therefore, a dual price system. When market control by producers is well established, prices may depart rather widely from those determined on commodity exchanges. However, producers with little market control frequently adjust prices to conform to those quoted on the commodity exchanges, or sell at the price quoted on a commodity exchange at the time of delivery.

This is well illustrated by the market for copper. Prior to the mid-1970s, domestic producers maintained substantial control over the U.S.-Canadian copper market and their prices, which tended to move in unison, were often considerably above or below COMEX and LME copper prices (see table 4-1). Since then, however, competition from outside producers has substantially increased. These outside producers, largely in developing countries, base their prices on those of the LME. As a consequence, U.S.-Canadian producer prices have recently been adjusted more frequently and stayed more closely in line with the commodity exchanges.

Prior to 1981 nickel was subject to strong market control by a few large Canadian companies. The producer price during 1982 and the first half of 1983 remained within 5 percent of the producer price in 1980 and 1981, while the LME price was about $1 per pound lower in both years. However, large discounts from the producer price offered by producers during the 1982–83 recession resulted in few sales at the published producer price. The Canadian producer price system was abandoned in late 1983 and nickel prices quoted by Canadian producers now follow those on the LME and the open market in the United States. Much stronger market control has been maintained by chromite ore producers throughout the 1970s and the first half of the 1980s. Between 1978 and 1984, producer prices of both South African and Turkish chromite have varied by only about 5 percent, despite the sharp decline in the demand for chromite in 1982–83. South African producer prices are set by a few large producers, while the Turkish price is subject to government control.

If a high proportion of a metal is produced by state enterprises in

TABLE 4-1. Copper Prices, 1960–82, in Cents per Pound

Year	U.S. Producer Price[a]	LME Price[b]
1960	32.0	30.7
1961	29.9	28.6
1962	30.6	29.6
1963	30.6	29.3
1964	32.0	43.9
1965	35.0	58.7
1966	36.2	69.1
1967	38.2	51.2
1968	41.8	56.1
1969	47.5	66.3
1970	57.7	63.9
1971	51.4	49.3
1972	50.6	48.5
1973	58.9	80.8
1974	76.6	93.1
1975	63.5	56.1
1976	68.8	64.0
1977	65.8	59.4
1978	65.6	61.8
1979	92.3	90.1
1980	101.4	99.3
1981	83.7	79.5
1982	72.9	67.2

Sources: Copper Studies, May 15, 1981, p. 1 for 1960–80; *CIPEC Quarterly Review*, April–June 1983, p. 90 for 1981 and 1982.

[a]annual average cash wirebar price

[b]annual average cash settlement wirebar price

developing countries, that metal's price tends to fluctuate widely because state enterprises characteristically do not cut back production in response to reduced demand in order to maintain prices. Yet this will only be true if there is a substantial number of producers, as is the case with copper. If the world market is dominated by only a few state enterprises, as with cobalt, producers are likely to determine most prices. The price volatility of a mineral may also be affected by an international commodity agreement or by collusion among producers—both of these influence the world price of tin, for example.

Another factor in price volatility is the existence of close substitutes for the metal. Copper, lead, tin, and zinc all have substitutes for a number of uses, but substitution takes place slowly in response to changing prices for metals or their substitutes, or to technological developments favoring substitutes.

The relationship between market characteristics discussed above and price volatility is shown in table 4-2. The correspondences are imperfect. For one thing, some market characteristics are difficult to assess. There are several prices for each metal at any one time. Market characteristics in major producing countries, such as the United States, may differ from those in the rest of the world, and international price movements are only partially reflected in national prices because of oligopolies and government import restrictions. A full understanding of price behavior, moreover, requires that one go beyond the relationships established in table 4-2 to undertake a comprehensive analysis of the market *structure* and special factors shaping price movements. Appendix 4-2 gives descriptions of the market structure of major minerals.

Short-Term Price Fluctuations in Mineral Markets

Table 4-3 shows annual fluctuations in prices of major minerals in current dollars over the 1978–82 period, but the picture of price volatility is incomplete for two reasons. First, quotations for all markets and quantities sold at those prices are not available, which means weighted averages of all prices for particular minerals cannot be computed. Second, one cannot distinguish between price changes peculiar to the specific market and those that arise from changes in the economy generally. The 1978–82 period was marked by a high rate of increase in all industrial prices (whose composite index rose nearly 50 percent). Price stability in *constant* price would exist if the annual increases in *current* price for a mineral matched the general index's rise.

However, very few industrial prices move precisely in tandem with the average of all industrial prices. Prices for each mineral move in response to the forces governing its own market. These include changes in price for competing minerals or for the commodities and services needed to produce

TABLE 4-2. Market Characteristics for Selected Minerals

Commodity	Degree of Vertical Integration[a]		Degree of Concentration in Production		Importance of Foreign Government Ownership	Trade on Commodity Exchanges	Existence of Close Substitutes	Price Volatility
	U.S.	International	U.S.	International				
Bauxite	high	high	n.s.	mod	mod	no	no	low
Alumina	high	high	high	high	mod	no	no	low
Aluminum	high	high	high	high	low	LME/COMEX	yes	low
Iron ore	high	mod	high	mod	high	no	no	mod
Steel	high	mod	high	mod	high	no	yes	low
Copper	high	mod	high	low	high	LME/COMEX	yes	high
Lead	high	mod	high	low	low	LME	yes	high
Tin	n.s.	mod	n.s.	mod	high	LME/Penang	yes	mod
Zinc	high	mod	high	high	low	LME	yes	high
Alloy metals								
Chromium	n.s.	high	n.s.	high	low	no	no	low
Cobalt	n.s.	high	n.s.	high	high	no	no	mod
Manganese	n.s.	high	n.s.	high	mod	no	no	low
Molybdenum	high	mod	high	mod	low	no	no	low
Nickel	n.s.	high	n.s.	high	low	LME	no	low

Notes: n.s. = no significant mine production; mod = moderate

[a]Degree of vertical integration refers to the control by a single company of several stages of mineral production from raw ore to refined metal.

TABLE 4-3. Average Annual Prices of Major Metals, 1978–82, in U.S. Dollars

Commodity	1978	1979	1980	1981	1982
Aluminum (lbs) producer price, ingot	0.54	0.61	0.72	0.76	0.76
Chromium (per mt chromite)					
South African producer price	56.00	56.00	55.00	55.00	52.00
Cobalt ($ per lb) producer price for metal	6.40–20.00	20.00–25.00	25.00	17.26–25.00	8.56 (average)
Copper (lbs)					
U.S. producer wirebar	0.67	0.93	1.02	0.85	0.73
LME	0.62	0.90	0.99	0.79	0.67
Iron ore (lt) Lake Superior ores basis 51.5% Fe	22.30–22.55	24.56–25.00	28.50–28.75	32.25–32.78	32.25–32.78
Steel (lbs) U.S. composite producer price	0.18	0.20	0.22	0.24	0.25
Manganese ore ($ per lb) 46–48% per lb, f.o.b. U.S. ports	1.28–1.42	1.36–1.42	1.38–1.75	1.66–1.76	1.58 (average)
Nickel ($ per lb) Canadian producer price	2.03–2.08	1.93–3.24	3.20–3.45	3.45–3.20	3.20
Lead (lbs)					
U.S.	0.34	0.53	0.43	0.37	0.26
LME	0.30	0.55	0.41	0.33	0.25
Molybdenum (lbs) U.S. producer price in concentrates	4.95	7.50	9.70	8.50	4.00
Tin (lbs) Penang market	5.68	6.72	7.46	6.38	5.87
Zinc (lbs)					
U.S. high-grade metal	0.31	0.37	0.37	0.45	0.39
LME, prime Western equivalent	0.27	0.34	0.35	0.39	0.34
U.S. index of producer prices for all industrial commodities	100	113	131	145	149

Source: Bureau of Mines, *Mineral Commodity Summaries 1983* (Washington, D.C.: U.S. Department of Interior, 1983).

TABLE 4-4. Average Percentage Deviation of Real Prices from Five-Year Moving Average, 1955–81

Mineral	Percentage	Mineral	Percentage
Copper	15.3	Iron ore	5.7
Lead	14.7	Bauxite	5.6
Zinc	14.4	Nickel	4.6
Tin	8.1	Aluminum	4.6
Manganese	7.2		

Source: World Bank, *The Outlook for Primary Commodities* (Washington, D.C.: World Bank, 1983), p. 45.

the mineral in question. Moreover, these influences often affect the mineral's price only after a considerable time lag. For none of the metals in table 4-3 has the average annual price moved proportionately to changes in the U.S. index of producer prices; in some cases (chromium and lead, e.g.) movement has even gone in the opposite direction. The metals also varied considerably in price stability.

One measure of commodity price variation is the divergence of real prices from a moving average over a given time period. Table 4-4 shows the average percentage deviation from a five-year moving average (in constant 1981 U.S. dollars) for nine major minerals between 1955 and 1981. The high real price variations for copper, lead, and zinc are due largely to their widespread international production and to the fact that all three commodities are traded on one or more of the commodity exchanges. The low fluctuations for aluminum, bauxite, and nickel reflect the high degree of concentration and vertical integration in their production. The low degree of variation in iron ore and steel prices results from the same kind of vertical integration and concentration, but here it is typical of steel production in developed countries. (Until recently, steel production in developing countries was for domestic markets, while developed-country producers controlled international markets.)

Long-Term Trends in Real Prices

Figure 4-1 plots movements by the combined price index of ten important nonfuel minerals for the 1950–85 period in both current and constant U.S. dollars. The index displays both cyclical movements roughly corresponding to changes in world business conditions and an overall downward trend. The latter results in part from a reduction in real costs of production, but it also reflects changes in market structures and production overcapacities for most minerals during the 1970s and early 1980s. These overcapacities existed because of the sharp decline in world metals consumption rates of growth

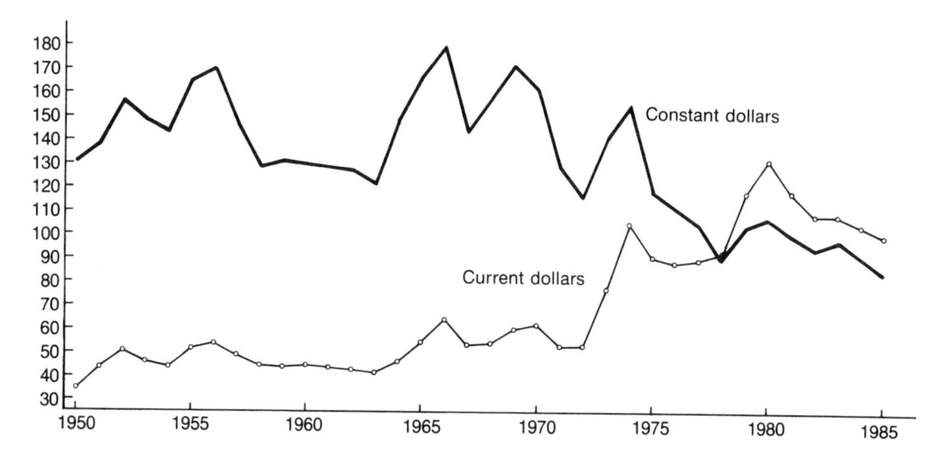

Fig. 4-1. Movements of combined weighted index of nonfuel mineral prices in current and constant dollars, 1950–84 (1977–79 = 100, annual averages). The commodities in the index weighted by 1977–79 export values are: copper, tin, nickel, bauxite, aluminum, iron ore, manganese ore, lead, zinc, and phosphate rock. (*Source:* World Bank, *The Outlook for Primary Commodities*, Commodity Staff Working Papers, No. 11 [Washington, D.C.: World Bank, 1984]; updated by author.)

beginning in the mid-1970s and because of the delayed effect of mining investments in Third World countries, largely by multinational mining companies, during the 1960s and 1970s. The downtrend, particularly since the mid-1970s, is also attributable to changes in world market structure. Multinational companies lost control of the markets due to competition from producers that did not reduce production in response to the decrease in growth of world demand. The policies of LDC state mining enterprises were partly to blame. But even where multinational mining companies retained ownership of mines in Third World countries, they were under pressure from governments to maintain output.

Thus overcapacity and changes in the world market structure, not just lower production costs, account for the long-term downtrend in nonfuel mineral prices. To be sure, the discovery of new reserves and advances in technology help to offset higher production costs occasioned by depletion of higher-grade ores. Even so, in 1980–84 minerals were significantly underpriced in comparison to their real costs.

For example, mining industry capital equipment costs have increased much faster than the wholesale manufacturing index. Wherever serious antipollution legislation is in effect, new capital costs arise. And the real cost of energy—especially important in aluminum production, but also for copper, lead, nickel, and zinc—is expected to rise later in this century. If the long-run growth in demand for minerals is to be accommodated, producers' increased costs for exploration, equipment, plant, and energy will have to be recovered through higher real prices.[6]

During 1982–84 real prices of five major metals (aluminum, copper, iron ore, lead, and zinc) were at their lowest levels since 1961 or before, and in 1985 the real price of tin was added to that list. These data, together with the low level of mineral industry investment in recent years and large increases in capital and other costs, suggest that strong world industry recovery would probably mean a sharp rise in nonfuel mineral prices during the 1990s.

Conclusions

The major topics discussed in this chapter—market structure for minerals, price volatility, and long-run price trends—all have important effects on investment in capacity. The market structure for a mineral influences its price volatility in response to changes in demand. Price volatility affects the timing of investment in new capacity. Unstable prices make for irregular investment in additional capacity and in the replacement or modernization of old capacity. Exploration outlays are also sensitive to sharp price movements. In part this is because current prices and profits influence expectations of earnings on new investments. This is true despite the fact that investments in new capacity will not result in increased output for five to ten years—when prices are likely to be quite different. This erratic pattern of investment decisions makes for continual imbalances between capacity and consumer demand, which in turn tend to exacerbate price swings. However, when there is a considerable degree of market control, prices are less ephemeral because output and sales are adjusted to cyclical shifts in demand while producer prices are kept reasonably stable. But if the market structure is such that a sizable portion of output is marketed regardless of price, prices will be highly volatile over the business cycle.

These short-term movements must be distinguished from long-term downward trends in real prices caused by changes in the growth of demand. Most investments in mineral exploration, mining, and processing capacity are made on the expectation that demand will continue to grow (on the average) at historical rates in future decades. If long-term rates of growth in demand fall, then investment in new capacity will result in overcapacity five or ten years later when the addition comes on stream. Declining secular rates of growth are therefore likely to bring on a long period of low prices and low profits (or even losses)—and consequently low levels of investment—until demand catches up to capacity. During the past decade such declines occurred for most metals, including aluminum, copper, steel, and the alloy materials used to make steel, such as nickel and manganese. The result has been overcapacity and low levels of investment. However, the fall in prices relative to costs differs among these metals depending upon their market structure. Given a substantial degree of market control, producers can minimize price decreases in spite of overcapacity.

The severity of an imbalance between capacity and demand will be greater following a long period of low prices and profits than following a cyclical fall in prices. In the latter case, investment declines less and for a shorter period. After a long time of low prices in relation to costs, investments will have been minimal for years and some capacity will have been abandoned. A strong world economic expansion following a long period of little or no capacity expansion might well result in a serious imbalance lasting several years. An industry like aluminum, characterized by vertical integration, may adjust to a long-run drop in growth of demand by maintaining prices and profits while setting its investment at each stage of production according to the lower rate of growth. By contrast, for copper, whose price has fallen well below costs in the United States, a sharp rise in demand would make both a sizable increase in prices and in U.S. dependence on imports.

Appendix 4-1

Summary of Market Arrangements for Nonfuel Minerals

The market arrangements discussed here cannot be organized into airtight categories. However, they may be classified roughly as follows.

1. Ores and concentrates sold on a contract basis for which open competitive markets do not exist because of substantial quality differences, high transportation costs, and special interrelationships between producers and processors. Examples include iron ore, bauxite, alumina, copper ores and concentrates, blister copper,[7] and lead and zinc ores and concentrates.

2. Metals produced in large part by firms that are integrated from ores through fabricated or manufactured products. The principal examples are aluminum and raw and unfinished steel.

3. Refined nonferrous metals, such as copper, lead, tin, and zinc, that are sold largely on a contract basis at producer prices or at prices governed by quotations on the commodity exchanges. The degree of producer control over prices differs from commodity to commodity and from country to country. There are also merchant markets for these metals.

4. Metals used mainly to produce ferrous alloys, including chromium, cobalt, nickel, and manganese, that are governed both by producer and merchant markets. Since there are relatively few producers, there is a large degree of market control over producer prices. However, since consumers are mainly steel producers, the number of buyers is small. There are merchant markets for these metals, but only nickel is sold on a commodity exchange.

5. Metal scrap, including copper, iron and steel, aluminum, and lead. Scrap competes with ore in metal production and in some cases constitutes a high proportion of the raw material used. Secondary copper, which is made entirely from scrap, competes in many uses with primary copper (produced mainly from ore), while steel made by electric furnaces employing scrap materials competes with the major companies that produce steel from iron ore. These products are sold outside the producer market dominated by large primary copper and steel producers. Merchant markets play an important role in this category.

Appendix 4-2

Market Structure and Pricing for Major Minerals

The following descriptions are designed to illustrate the characteristics of mineral markets discussed in the text. No attempt is made to analyze the competitiveness of these markets or their relationship to investment and availability of mineral supplies. Such an analysis would require a book in itself. The accounts of aluminum, copper, and steel markets help to understand the changing competitive position of U.S. producers of these metals as discussed in chapter 5. Understanding the markets in metals for which the United States is largely import dependent is important to the discussion of our vulnerability to import disruption in chapter 5.

Bauxite-Alumina-Aluminum[8]

In 1978, six large integrated firms owned or had an equity interest in over half the aluminum capacity of the market economies; about fifty other private companies owned a quarter of capacity, frequently in association with one or more of the six large firms; and twenty-four governments owned or had an equity interest in the remaining capacity. The major firms are (1) Aluminum Company of America (ALCOA, United States); (2) Alcan Aluminum Ltd. (Canada); (3) Reynolds Metals (United States); (4) Kaiser Metals (United States); (5) Pechiney Ugine Kuhlmann (PUK, France); and (6) Alusuisse (Switzerland).[9] As of 1980 these six companies controlled 45.0 percent of the bauxite capacity of market economy countries; 62.5 percent of the alumina capacity; and 52.5 percent of the aluminum capacity. The Soviet Bloc countries controlled 16.5 percent of world aluminum capacity and China 1.8 percent.[10]

Although there is a fairly high degree of concentration in the industry, it has decreased significantly since 1958, when the six firms controlled some 80 percent of aluminum capacity outside the Communist countries and somewhat higher percentages for bauxite and alumina. The largest expansion in capacity since 1958 has been in firms wholly or partially owned by governments, including those of Bahrain, Brazil, India, Italy, Norway, South Africa, Spain, Turkey, and Yugoslavia. A number of state enterprises are planning new aluminum capacity, including Brazil, Trinidad, and Venezuela.[11] Governments have also increased their interest in production of bauxite and alumina, particularly in bauxite-producing countries.[12]

The six largest aluminum companies are more or less self-sufficient in both bauxite and alumina. These companies also fabricate most of the aluminum they produce. ALCOA, Reynolds, Kaiser, and Alusuisse fabricate over 90 percent of their aluminum output, while Alcan and PUK have fabricating capacity equal to 70 to 90 percent of their aluminum smelting capacity. The so-called second tier aluminum producers (Revere, National Southwire, Martin Marietta, Ardal og Sunndal Verk [ASV], and British Aluminum) have only minor bauxite holdings, but in the aggregate produce over half their alumina requirements. Except for ASV, this group also produces the bulk of its own aluminum. Another group of private companies whose primary business is in products other than aluminum (AMAX, Anaconda, Billiton, Noranda, and Phelps Dodge) accounts for nearly 23 percent of bauxite production of the market economies, 11 percent of alumina, and nearly 8 percent of aluminum. Some of them, including AMAX, Anaconda, and Phelps Dodge, fabricate a high proportion of their aluminum production.[13]

Because of the high degree of vertical integration in the bauxite-alumina-aluminum industry, an estimated 85 to 90 percent of the bauxite produced in the market economies is consumed by the enterprises or consortia producing the bauxite, or by associated enterprises. Most of the remainder is sold on a contract basis to independent processors, and there is only a small amount left for spot transactions, most of which are handled by merchants. Likewise, an estimated 80 to 85 percent of alumina production is consumed by the producing enterprises or their associates. Although there is no visible spot (or terminal) market for alumina, the level of merchant activity in alumina is somewhat higher than for bauxite. Most aluminum producers maintain posted prices in individual country markets, but about 10 percent of the world's aluminum trade takes place in the merchant market.[14]

In late 1978 trading in aluminum metal began on the LME, followed in 1983 by the COMEX. The LME price varies much more than producer prices. The LME provides a hedge for producers and consumers, as well as a market for firms with excess metal. In 1980 physical deliveries on the LME were estimated at about 0.5 percent of world production,[15] although trade there has been rising. Thus far, LME quotations appear to have little influence on producer prices.

Despite the high degree of concentration and the existence of price leadership in the aluminum industry, it experiences substantial competition. This is indicated by the entry of new firms over the past two decades and the decline in the real price of aluminum since the late 1960s despite sharp rises in energy costs. As more aluminum becomes available in international markets, market control by major producers is likely to weaken. If so, as with copper, producer prices will more closely approximate open-market prices like those on the LME.

Iron Ore

In 1977 approximately 85 percent of U.S. iron ore output was produced from eighteen mines operated by eight companies, and an estimated 82 percent of the ore output was produced by or for the account of U.S. companies engaged in the production of iron and steel; 3 percent was for the account of Canadian steel companies; and 15 percent was owned by independent producers.[16] The U.S. imported 30 percent of its iron ore requirements during 1980–83, about two-thirds of which came from Canada (largely from mines owned by U.S. steel producers). In Western Europe about one-half the iron ore comes from captive mines (i.e., mines owned or controlled by steel companies), with the remainder sold under long- and short-term contracts, or on a spot basis. Japan acquires all its iron ore abroad, much of it on long-term contracts. There is a relatively small amount of U.S. and Canadian ore sold on a spot basis, mainly by merchants, and some is traded under short-term (less than a year) contracts at a fixed price. In the United States and Canada, the large steel companies and leading merchants announce prices on an Fe grade (iron content) basis according to a formula for major types of ore, including pellets (iron ore concentrated into balls), the major form in which it is shipped.[17]

Contracts govern most internationally traded ore, and these vary considerably in terms and in pricing methods. One type of long-term contract helps to underwrite mine development projects, with consumers supplying a portion of the financing. Initially some of the long-term contracts provided for fixed prices, but rising inflation in the 1970s made them untenable for producers. More recent contracts have provided for periodic renegotiation of prices or automatic adjustments to inflation. Long-term contracts provide an assured market for producers and guaranteed supplies for consumers. Japanese companies have negotiated a number of such contracts with Australian and Brazilian firms, and in some cases Japanese concerns or their affiliates have contributed to the financing of iron ore projects. Iron ore, iron ingots, and steel are not sold on commodity exchanges.

U.S. Steel Industry Prices

The U.S. steel industry is composed of approximately 180 companies, which produce a wide range of products. Of these companies, 86 have their own steel-making furnaces, while the others purchase semifinished steel for fabrication. In 1977, 19 companies were fully integrated from iron ore to blast furnaces, to steel-making furnaces, to finishing mills; and 40 operated 51 minimills (electric-arc furnace plants that each produce 600,000 tons or less of raw steel annually). There are dozens of iron and steel products, from raw steel to various alloy steels with special characteristics required for industrial uses.[18] Some of these products are produced by companies that are neither integrated steel firms nor minimills.

U.S. integrated steel companies sell on the basis of producer prices and adjust output to changes in demand. Under normal demand conditions, producer prices set by individual firms for various products tend to be more or less uniform due to price leadership. However, in a period of abnormally low demand such as that in recent years, the integrated steel companies sell at large discounts from the official producer prices. Minimills, on the other hand, adjust their prices to maintain output at more or less full capacity, so their prices are competitive with those of imports as well as with those of integrated steel firms.

Both imports and U.S. government import controls affect prices for steel products in the United States. Import prices reflect marketing policies of foreign producers selling in the U.S. market; thus Japanese, Australian, and European prices may differ. In 1973–75, a period of strong demand, imported carbon steel products sold for more than competing U.S.-produced products, but when the boom subsided in 1975, prices for imported products became lower than domestic prices. Despite the oligopolistic nature of the integrated U.S. steel industry, imports plus minimills and other nonintegrated firms tend to make the industry basically competitive.[19]

Foreign Steel Pricing

The major steel-producing and -exporting industries outside the United States are subject to a variety of government controls, subsidies, and cartel activities, and there is substantial government ownership of steel firms in other market-economy countries. In the EEC countries the European Coal and Steel Community (ECSC) is empowered to fix maximum or minimum prices, to establish production quotas or allocate steel production, and to sanction national cartel-like practices. The Japanese iron and steel market is highly oligopolistic, and Japanese law recognizes domestic cartels. Export cartels are also permitted, and they govern export trade in a number of products.[20] There is sizable government ownership of steel production in the United Kingdom, Austria, Ireland, France, Italy, the Netherlands, Norway, Portugal, Spain, and Sweden; among developing countries, government ownership of steel production prevails in Argentina, Brazil, Colombia, India, and Mexico, among others.[21] Government ownership reduces or eliminates competition in the domestic market by means of price, production, and investment controls, even where private firms also operate. It often means that exports are subsidized. Export prices are usually determined by market conditions in the countries where foreign steel companies sell and may bear little relation to prices in the exporting countries themselves.

Copper

Copper is traded at different stages of processing, from ores to concentrates, to blister (produced by smelters before refining), and finally to refined copper. Primary refined copper is produced largely from copper ore, while secondary refined copper is produced from scrap. There is a large market for many types and grades of copper scrap. Owing to the high degree of vertical integration in primary copper production, most copper traded here and abroad is refined. There are several standard types of primary refined copper, including wirebar, cathode, and continuous cast rod (CCR).[22] Quality differences among the standard types give rise to price differences. Primary and secondary copper are competitive in certain uses. There is also a substantial volume of trade in copper alloy products.[23]

The U.S. industry is highly concentrated, while the world copper industry is moderately concentrated. In the United States, which normally produces about one-fourth of the copper mined in the market economies, ten companies control over 90 percent of total mine capacity and the three largest producers—Kennecott, Phelps Dodge, and Newmont—account for 50 percent of mine capacity. Three foreign government mining enterprises control one-third of the copper mine capacity in the market economies;[24] approximately 20 percent is controlled by a dozen U.S. and foreign multinationals producing outside the United States (see chap. 3); and most of the remainder is controlled by private concerns operating in the country of production. The principal U.S. mining companies with considerable ownership or control of foreign production include AMAX, ASARCO, Freeport-McMoRan, Newmont, and Superior Oil. Foreign mine copper production controlled by U.S. companies represents about 13 percent of all market-economy output.

Most large U.S. primary copper producers have their own smelters, but over three-fourths of the U.S. primary refining capacity is held by only three firms, namely, Kennecott, Phelps Dodge, and ASARCO.[25] U.S. mining concerns without smelting or refining facilities have their raw material processed either under toll contracts (involving payment for processing, following which the refined copper may be marketed by the mining company) or by direct sale of the materials to smelters or refineries. The smelters and refineries also process concentrates from Canadian, Philippine, and Latin American mines. Recently some U.S. copper concentrates have been processed in Japan and Germany.

Although large government-owned mining enterprises in some developing countries, including Chile, Peru, Zaire, and Zambia, smelt and refine most of their own mine output, a substantial amount of concentrates from both government and privately owned foreign mines is sold to firms in Japan, Western Europe, and the United States with smelting and refining capacity. During the 1964–74 period there was a large increase in the supply of concentrates from nonintegrated mines, particularly those in Canada, the Philippines, Papua New Guinea, and Indonesia. Accompanying this was a rapid growth of smelter capacity in Japan and West Germany, the largest importers of copper concentrates. These concentrates are usually sold under long-term contracts, with prices determined by LME quotations.

Although the domestic markets for refined copper in the United States, Western Europe, and Japan each have unique features, the copper market is basically global. Thus LME prices tend to move parallel to those on the COMEX, and prices on metal exchanges influence producer prices. Since 1977 producer prices quoted by major integrated U.S. primary copper producers—Kennecott, Phelps Dodge, ASARCO, and Newmont—have followed COMEX and LME price movements more closely than they did in previous years. Copper is also bought and sold by merchants who deal in various grades of domestic and foreign refined copper, and in scrap. Merchant prices are usually based on metal exchange quotations.

Price leadership has kept producer prices of major North American primary producers more or less in line with one another for the same grade of copper. In the past, the U.S. producer price (average of individual producer prices) often diverged sharply from LME or COMEX quotations (see table 4-1).[26] Between 1964 and 1974 U.S. producer prices for wirebar were in most years substantially below LME prices, but they were significantly above LME quotations following the sharp fall in copper prices in 1974. However, since 1977, increased competition from foreign producers (who tend to base their prices on LME quotations) has helped narrow the margin between U.S. producer prices and those quoted on commodity exchanges.

Other Nonferrous Metals

Lead

Lead is produced by a number of countries but mine production, and to a greater extent refined output, is concentrated in developed countries—the United States, Australia, and Canada for mine production, and these same countries plus the EEC countries and Japan for refined output. As with copper, North American lead producers sell refined lead and buy ore concentrates from nonintegrated mines on the basis of producer prices, while outside North America prices are based on LME quotations. Concentrates compete heavily with lead scrap, which constitutes over 40 percent of lead consumption in the United States. The U.S. producer price was higher than the LME price during the 1977–84 period (except for 1979), but the differential was usually less than 10 percent.

Zinc

Some eighty zinc mines operating in the market economies are controlled by twenty-five companies that produce about 85 percent of mine output. These concerns are located mainly in Canada, the United States, Australia, Peru, and Mexico. Eight companies control more than 50 percent of the market economies' zinc mine capacity, and four Canadian firms control about one-third of capacity. However, over half of Canada's mine output is smelted abroad. U.S.-based companies control about a third of mine production in the market economies. Foreign interests own the major Australian and Mexican mines, but only 20 percent of Peru's mine capacity is under the control of foreign companies. The zinc smelting industry in the market economies is more concentrated than that of zinc mining. There is a high degree of vertical integration in zinc, with many large smelting companies integrated forward to metal manufacturing. In 1974 the United States had 600,000 tons of zinc smelting capacity, all of which was controlled by six corporate groups that also controlled about 50 percent of U.S. mine production. Between 1974 and 1983 there was a substantial reduction in U.S. zinc smelting capacity, due mainly to obsolescence and to the high costs of legally required pollution abatement.[27]

As with copper and lead, U.S. firms sell on the basis of U.S. and Canadian producer prices. However, foreign zinc smelters use a European producer price called the Commonwealth Producers Price (CWP). Both the U.S. producer price and the CWP are used for pricing ore concentrates sold to smelters by nonintegrated firms as well as for sales of refined metal. The U.S. producer price tends to follow the CWP price, but both are influenced by LME quotations.[28] Despite the sharp fall in world consumption of zinc during the 1981–82 recession, zinc prices declined much less than those for copper and lead, suggesting a greater degree of market control by producers. Capacity utilization of U.S. primary zinc smelters was only about 50 percent at the end of 1982, a good indication of the producers' ability to maintain price leadership by cutting back production.

Nickel

In 1950 nickel production was the virtual monopoly of one company, International Nickel of Canada (now INCO), which accounted for 85 percent of total market-economy sales; the remainder was shared with Falconbridge (Canada) and Le Nickel–SLN (France) operating in New Caledonia. By the mid-1970s, however, INCO's share of both mine and refined nickel and ferronickel output had declined by more than half. The development of capacity in Australia, South Africa, the Philippines, Botswana, and other countries helped account for this. The Hanna Mining Company's Oregon mine supplied about 10 percent of U.S. requirements. There is a good deal of nickel-smelting and -refining capacity in Western Europe, Japan, and the United States, but nearly 30 percent of its total remains in Canada.

Most nickel is used in alloys, particularly in stainless and other steels. Nickel is marketed in several forms, including nickel metal and ferronickel (containing 40 to 50 percent nickel).[29] Prior to 1977 INCO was the price leader. In 1977 it abandoned this role, but a group of Canadian producers together maintained fairly uniform producer prices until 1983, when Canadian prices began to follow LME quotations and open market prices in New York. Nickel has been traded on the LME since 1979, and U.S. merchants conduct an active free market in the commodity.[30]

Tin

Most tin produced by market economies comes from Malaysia, Thailand, Indonesia, Boliva, and Australia. Malaysia, Thailand, and Indonesia have enough smelting capacity for their own mine production. Nonintegrated mining companies sell their tin concentrates to smelters under long-term contracts. The principal physical market is the Penang exchange in Malaysia, while the LME market is mainly in futures. Supply contract prices are usually based on the quotations of these two. For more than twenty years price has been strongly influenced by the International Tin Agreement (ITA). The ITA buffer stock attempts to keep prices within a certain range by means of sales and purchases. Since the U.S. strategic stockpile has excess tin inventories, the General Services Administration (GSA) occasionally sells tin; these sales have an important effect on the market. The governments of Malaysia, Indonesia, and Thailand also influence it by means of export controls.[31] As a consequence of these three factors, price fluctuations were moderate until late 1985, when the ITA buffer stock could no longer support tin prices within the desired range. Future developments will be greatly influenced by whether the ITA will be able to borrow enough funds to resume its former role.

Manganese

In 1981 over 40 percent of the manganese ore sold in market economies came from South Africa; Brazil, Australia, Gabon, and India produced the remainder. An increasing share of manganese exports takes the form of ferromanganese; South Africa, Norway, and France account for about 70 percent of market-economy exports.[32] About three-fourths of all manganese produced is used as ferromanganese to manufacture steel alloys. Both manganese ore and ferromanganese are purchased through annual contracts (between producers and consumers) or from merchants (who contract for supplies from producers). The largest concerns sell at a producer price, but small producers and merchants trade on a free-market basis at discounts or premiums to the producer price.[33] The price of manganese ore remained quite stable over 1975–81. Despite the sharp fall in steel production, the average price of manganese during 1982–84 was only 15 percent lower than in 1981, indicating a substantial degree of market control by producers.

Chromium

Among the market-economy countries, the main producers of chromite ore are South Africa, the Philippines, Zimbabwe, Turkey, Brazil, and India. The USSR and Albania are also producers and exporters, and the United States imports some chromite from both these countries (as well as from South Africa, the Philippines, and Turkey). An increasing portion of world trade in chromium is in ferrochromium, and here South Africa dominates production and export, followed by Yugoslavia, Zimbabwe, and Sweden. In 1980 the United States imported about 40 percent of its ferrochromium requirements and all its chromite ore. Chromium is used (primarily in the form of ferrochromium) to produce alloys, mainly stainless steel.[34]

Most chromite ore and ferrochromium is sold directly by producers and their agents to large steel companies and other users under long-term contracts at producer prices. South Africa tends to be the price leader for high-iron chromite; there is also a published price for Turkish high-chromium chromite, which is usually about double the South African price because of its higher chromium content. There is an active merchant market for smaller steel companies and foundries.[35] The South African producer price remained fairly constant during 1977–81 and fell by less than 10 percent in 1982–84 (average), despite a sharp decline in consumer demand. This indicates a high degree of market control by producers.

Cobalt

The most important producers of cobalt in the market economies are Zaire and Zambia, which accounted for about two-thirds of mine output in 1980 and over half the metal output. Some Zairian cobalt is processed in Belgium. Most cobalt is a by-product of either nickel or copper mining. As with chromium, the principal use of cobalt is for alloy steel, including "superalloys" used in the aircraft industry. Most of the world's trade is in metal or other processed forms.[36]

Cobalt is mainly sold under contract at producer prices directly to large steel concerns, to companies using it for other alloyed materials, or for chemical and ceramic uses. The price leader is SOZACOM, the marketing agent of Gecamines, Zaire's government-owned mining company.[37] There is also an active merchant market, where prices are more volatile. For example, the merchant-market price more than tripled the producer price for cobalt metal during the shortage in 1978–79 after the invasion of the Shaba province in Zaire. Due to that shortage, SOZACOM adopted an allocation system for shipments at the producer price; it rose from $6.40 per pound in early 1978 to $25.00 per pound in 1979. In 1980 the producer price remained at $25.00, dropping to $17.00 in 1981 and to $12.50 in 1982 (where it remained through 1984) in response to the slacking of world demand. However, the U.S. merchant price fell to $4.00 per pound toward the end of 1982, but rose to $11.00 in 1984. Despite high production concentration, market control by producers has been weak since 1978. This is due in part to the sharp decrease in demand, reflecting both substitution of other materials for cobalt and, in part, to the reduction in steel production. Cobalt output is difficult to control because it is produced as a by-product of other metals.

Other Metals

World markets for most other metals, including columbium (mainly Brazil), molybdenum (largely United States), and platinum-palladium (South Africa and the USSR), are similar to that for chromium, with most trade taking place between producers and large consumers at producer prices, but with free merchant-market prices above or below them in times of shortage and surplus, respectively.[38] An exception is tungsten, produced by a large number of countries, but whose market is mainly competitive.

5

Maximizing Nonfuel Minerals Supply: Free Trade versus Import Restriction

The United States has traditionally maintained a liberal foreign trade policy with respect to nonfuel minerals, even for those in which foreign producers were strong rivals for the domestic market. Tariffs on primary metals tend to be low.[1] Prior to the late 1970s nontariff barriers (ntbs) were rarely employed. Since then, however, this country has become more protectionist, both generally and with respect to nonfuel minerals, which has resulted in increased use of ntbs on steel products and strong support for ntbs on nonferrous metals.

The principal arguments for protection of domestic nonfuel minerals industries are promotion of national economic welfare, including job retention, and reduced foreign dependence for reasons of national security. Without further definition, these contentions amount to slogans rather than rational arguments. In an interdependent world economy, economic welfare is maximized by national specialization and trade in competitive international markets. Under these conditions consumers can buy in the cheapest markets; producers, for their part, maximize total value output by specializing in areas where their relative efficiency or comparative advantage is greatest. In a world economy without barriers to trade and to movements of capital, technology, and skills, productive factors will flow to areas with the richest resources; nonfuel minerals will be available at the lowest price and productive capacities will grow apace with world demand. Artificial barriers to trade in goods and productive factors reduce both national economic welfare and availability of nonfuel minerals.

Thus protection for nonfuel minerals producers must be shown to be justified either by U.S. vulnerability to import disruption or by the need for domestic capacity to supply essential military and other products in times of emergency. Basic mineral product industries such as steel might justify such favored treatment, because the national requirement for them is subject to changing specifications. Aluminum and copper, however, are among the metals that fail this test, since they are produced internationally in a few standard grades. Due to increased industrial use of minor metals like chromium, cobalt, and manganese, for which the United States depends almost entirely on imports, concern has also arisen about the threat this dependence could pose to economic viability, whether in time of peace or war.

The general argument for maintaining or increasing national self-sufficiency in basic minerals or mineral products, such as aluminum, copper, and iron and steel, might be called the "industrial base" rationale. It raises important questions that proponents have failed to answer. The first is why they single out only these commodities in the entire U.S. industrial base for a high degree of self-sufficiency. The economic welfare and defense capability of the country depends on bauxite to produce aluminum and iron ore to produce steel, but few if any voices urge that the United States reduce its import dependence on these commodities. More important, the country is increasingly dependent on imports for a wide range of industrial products, many of them needed for domestic civilian and military production. Yet no one has come forward to argue for absolute industrial autarky.

The second question is what degree of self-sufficiency in any particular product or industry will optimize national welfare and defense capability. Some argue, for example, that the latter goal makes a strong iron and steel industry mandatory. Does this mean the United States should be virtually autarkic in each of hundreds of iron and steel products? How much self-sufficiency in stainless or high-carbon steel is required?

A third question for which the answers have yet to be supplied is under what conditions imports of any commodity or category of commodities might be unavailable to the domestic economy. During an all-out war that made sea and air unsafe for international commerce? Or with disruption of specific imports from certain foreign countries because of political and economic conditions there?

A final question is whether the best remedy for vulnerability to import disruption lies in the effort to achieve self-sufficiency or in some other policy, such as stockpiling. If the imports come almost exclusively from a few unstable countries, disruption of supplies from any one of them could impair industrial production for both defense and civilian uses. But if imports of important products come from a number of countries, most of which are not subject to political or economic instability, disruption of imports from any one of them may cause prices to rise somewhat, but that would not seriously undermine American industry. Promoting self-sufficiency may often prove the more costly alternative. And it may not even be the most effective.

The Department of Defense gives preference to U.S. firms in purchasing a large number of military items, arguing that foreign suppliers are less reliable than domestic sources. This may be true for highly specialized products, but not necessarily for standardized materials. Because there are usually many foreign supplies of standardized products, domestic sources are not always the most reliable. For example, strikes in the United States may curtail production for a considerable time. In any case, reliance on the domestic economy must clearly have some limit. Otherwise, all defense industries would have to buy all components and raw materials in this country regardless of cost.

It is reasonable to argue that the U.S. iron and steel industry should be sufficiently large and broadly based to accommodate special defense requirements and to adapt quickly to changes in them. The question remains whether this applies to all ferrous products needed by the defense industry, and how much self-sufficiency each product requires. The latter would depend in part on what portion of total demand is represented by defense-industry consumption. If, for example, defense requirements constitute only 5 to 10 percent of total U.S. needs for a product and are unlikely to rise above that level except in an all-out war, a 30 percent overall domestic production of each would appear adequate to fulfill the special requirements of defense.

As for the minimum degree of autarky needed in more or less homogeneous products such as aluminum and copper, no special manufacturing capacity or adaptability is required. If all import sources are disrupted, perhaps minimum domestic capacities related to actual and potential defense requirements should be maintained. But, alternatively, the products in question might be stockpiled.

Many kinds of national emergency might threaten the availability of raw and processed materials and of intermediate and finished products to the United States. It would not be economically feasible to prepare for all possible contingencies, any more than it is militarily feasible for the Defense Department to prepare for every conceivable combination of war emergencies. Normal practice in such matters is to compare the estimated social costs of a particular contingency, adjusted for the probability of its occurrence, to the social costs of avoiding or mitigating it. There may also be an overall financial limit on measures that may be undertaken. The defense budget determined by Congress and the administration is an example of such constraint. Within the budget's limits, national security is maximized when an additional dollar of expenditure on any program yields the same security benefit as the last dollar expended on any other program. The same principle should apply to the supply of nonfuel minerals important to national security.

Maintaining autarky in basic mineral products to meet both civilian and defense requirements during a long period of all-out war would not be cost effective. First, the probability of a war similar to World War II is very low.[2] Second, under conditions of a long conventional war, the United States would employ only a fraction of the basic industrial metals it uses in peacetime to produce houses, autos, electrical appliances, office buildings, public works, civil aviation, and the like. Most of its manpower and industrial resources would be mobilized in war. The civilian economy would no longer command a large portion of the basic minerals this country normally produces or imports. Third, the social costs of virtual self-sufficiency in basic mineral products would be very high in terms of both foreign retaliation against import restrictions and the higher prices U.S. consumers would pay for these products. For some of them, such as copper and aluminum, stockpiling would be cheaper than achieving substantial autarky, although the variety and frequent changes in the kind of steel products makes stockpiling infeasible.

Most of this chapter analyzes the future U.S. competitive position in three important metal industries: copper, steel, and aluminum. Some fear that growing import dependence in these industries will weaken the nation's industrial base, with consequent harm to our defense capability and—in some vague way never satisfactorily explained—to our economic welfare. I do not think that U.S. economic welfare in peacetime requires any large measure of self-sufficiency for the nonfuel mineral industries. I do grant that a certain degree of self-sufficiency in these industries may be important for national defense. An assessment of these industries is therefore relevant to the larger policy issue of how to maximize availability of minerals to the economy in peace or in war. It seems only fair, however, to conclude this discussion with a brief account of the demonstrable and probable costs (even for national defense) of the proposed abandonment of free-trade policies. There is less than meets the eye in most arguments for increasing defense readiness by decreasing imports of nonfuel minerals.

Copper

In mid-1984 a leading U.S. mining official stated that without import restrictions on copper 60 percent of the domestic industry would be gone by 1990.[3] How accurate is this statement? What would happen to national security in the 1990s and thereafter if the prediction holds?

Nuclear attack aside, the most serious copper emergency would be several years of conventional war that ended ocean shipments to North America, limiting U.S. imports to those from Canada and Mexico. In 1984 these two countries' excess production over consumption constituted about 30 percent of U.S. refined copper consumption. Were Canada and Mexico to continue expanding their production in line with world demand, their excess capacity (which, in this scenario, could not travel by sea) would be able to provide about 30 percent of U.S. copper consumption. In addition, secondary production based on copper scrap supplies about 25 percent of U.S. refined copper consumption. Thus only 45 percent of America's normal copper requirements would have to come from U.S. primary output from mines, as contrasted with about 50 percent in 1984. Under wartime conditions effective civilian demand would be lower than in peacetime. Therefore, given the conventional war scenario, U.S. primary production could decline to 60 percent of its 1984 level and (together with imports from Canada and Mexico and domestic secondary production) the United States could still meet 85 percent of its 1984 copper requirements.[4]

One copper specialist believes that by 1990 only 30 percent of U.S. copper consumption will come from primary domestic ore production as contrasted with about 50 percent in 1984. This projection is based largely on the expectation that several large domestic copper smelters will be closed and will not be replaced by facilities capable of meeting EPA standards. Mine production will be somewhat higher, but almost 20 percent of U.S. mine output in the form of concentrates is expected to go abroad for treatment.[5] This forecast assumes that real copper prices will remain below costs for many mines, and that imports will rise from about 26 percent of consumption in 1984 to about 48 percent in 1990. There is reason to believe this forecast is too pessimistic. Some mining firms have reduced their costs by $0.20 per pound or more during 1984 and 1985, turning former losses into modest profits. World investment in new capacity will remain low compared to growth in demand until prices exceed real costs. This should bring about a rise in real copper prices sometime during the next decade. Finally, it is difficult to believe the United States will shift from near self-sufficiency in copper in 1981 to a 48 percent dependence on imports within less than a decade.

After World War II, U.S. copper mine production steadily increased, reaching a high of 1.55 million mt in 1973; after a decline in the second half of the 1970s, it rose again to about the same level in 1981, although as a percentage of non-Communist output, U.S. production dropped from 36 percent (1950) to about 25 percent (1979 and 1981). On balance, the United States relied upon foreign suppliers for about 6 percent of its 1981 copper needs, but it was potentially self-sufficient.

The relative decline in U.S. output parallels that of its copper reserves as a percentage of non-Communist world reserves—from 34 percent in 1950 to 22 percent in 1980.[6] Many other factors determine such a ratio to world production. It includes the quality of reserves, location of mines, government policies on investment in mining, and the relative costs of such inputs as labor and energy. Copper mined in the United States is about 40 percent below the world average grade, and labor costs here are the highest in the world. All the same, a favorable investment climate, skilled personnel, and superior infrastructure in the mine areas continue to make developed countries attractive to investment from abroad.

During the 1982–83 recession, U.S. and world consumption of copper declined sharply. The consequent low copper prices, mine shutdowns, and reduced mine exploitation meant a sizable decline in U.S. output: from about 1.5 million mt in 1981 to about 1.0 million mt in 1983. Some of the shut down mines are unlikely to be reopened. All the same, U.S. *potential* output at the end of 1983 was estimated at about 1.33 million mt.[7] Before output could approach this magnitude, of course, copper prices might have to go as high as $1.00 per pound. (The mid-1984 level was less than $0.70 per pound [U.S. producer price].)

Copper prices will probably rise enough in the next few years to restore a large portion of U.S. mining production before 1995.[8] Even if there were no increase in capacity during the remainder of the 1980s, and even if the 1980s average of 2 percent per annum growth remained constant, U.S. copper-producing capacity, including secondary production, would still equal 65 to 70 percent of U.S. consumption in the early 1990s. This will require the building of new smelters or the installation of leach-electrowinning plants in place of smelters. In 1986, Phelps Dodge announced plans to expand its smelter capacity in Arizona and Kennecott was in the process of modernizing and expanding its mine and smelter facilities in Utah.

A significant addition to U.S. capacity might require a copper price of $1.30 per pound (in 1980 U.S. dollars).[9] A number of factors will decide whether that price will prevail. They include the world rate of growth in demand and the expansion of copper capacity in other countries, particularly in South America and in the Asia-Pacific region, where production costs are lower, and the investment climate often less promising, than in the United States (see chap. 3).

Whether U.S. copper-producing capacity expands or declines will also depend on U.S. production costs relative to those abroad. Relative costs are difficult to forecast because of foreign exchange fluctuations, pollution-abatement expenses, and the introduction of cost-reducing technology, but a comparative study in the copper industry found that 1978 U.S. production costs were 20 percent above world average (see app. 5-1). Moreover, CODELCO (Chile), the non-Communist world's largest copper producer, turned a profit in 1982–83 when nearly all U.S. (and most foreign) producers were incurring losses.[10]

Although U.S. average costs are likely to remain higher than world average costs, the domestic copper industry is not likely to disappear by 2000 or soon thereafter. Individual U.S. copper mine costs range from below the world average to well above it, and this is also true for individual mines abroad. Moreover, ample financial resources for investment in new capacity, and a more favorable economic and political climate than in most other countries, gives the United States an important edge. Although its competitive position will probably decline, the United States is likely to be at least 50 percent self-sufficient in copper during the remainder of this century.

The 1982–84 crisis was mainly a consequence of temporary factors; several copper specialists believe that within this decade prices will again rise enough to restore most of the cutback in production except for permanently closed mines.[11] The Canadian mining industry, one of the lowest cost in the world, cut back mine production about 17 percent between 1980 and 1982–

83. In 1981 U.S. mine production was within 2 percent of its 1973 all-time peak, and refined copper output was within 7 percent of its 1973 peak. Moreover, U.S. mine production as a percentage of non-Communist world production remained fairly constant—about 23 percent—during the 1970s. Large industries do not expire as a result of short-term crises; they decline slowly over long periods from fundamental factors.

Steel

The U.S. steel industry, which dominated the world market for over seventy years, began losing its international competitive position in the 1950s and has lost ground ever since. Net iron and steel imports grew rapidly during the 1960s, whereas in 1957 exports had balanced imports. Net import reliance rose from 8 percent in 1976 to 12 percent in 1979; the rise continued during the early 1980s, to 22 percent in 1982 and to an estimated 25 percent in the first quarter of 1984.[12] In contrast to copper, raw steel production has fallen since 1973. By 1980 it was only 74 percent of the 1973 figure, and less than 50 percent by 1982. The steel as well as the copper industry has suffered from a combination of reduced growth in world demand, substantial world overcapacity, and a cyclical reduction in demand.

Just as with the copper industry, cyclical and transitory factors that diminish the U.S. steel industry's competitive position should be distinguished from longer-run structural causes. Transitory factors include the 1982–83 recession that worsened the previous world overcapacity, leading to strong foreign competition for the U.S. market; and the 30 percent appreciation of the dollar against other major currencies between 1980 and early 1984. More significant for the long run are structural factors operating in the 1970s, some of which may remain important in the future. A major example is global steel production overcapacity, resulting from the slowdown in the growth rate of steel consumption worldwide between 1949 and 1979.[13] During the 1969–79 period foreign steel capacity, particularly in Japan, the EEC, and in LDCs, grew much faster than steel consumption, while domestic capacity grew very slowly. This disparity helps to explain the higher rates of modernization and adoption of new technology by major foreign competitors of the United States. A second, longer-term impairment of the U.S. competitive position arises from high wages, which exceed those in other manufacturing industries by a large margin. (These and other structural factors are detailed in app. 5-2.)

Outlook for the U.S. Steel Industry

In 1983 a major restructuring began when large integrated steel companies closed older, high-cost facilities. For instance, U.S. Steel announced it would shut down three plants and portions of others and eliminate over 15,000 jobs.[14] Recent mergers, such as that between Jones and Laughlin (a subsidiary of LTV Corporation) and Republic Steel, were accompanied by announcements that some old plants would be closed and others modernized.[15] A leading steel economist estimates that the eight largest integrated firms will reduce their capacity from 138 million short tons (st) in 1983 to 105 million st in 1985, with still further reductions by 1990.[16] Lower costs and the introduction of new technology are expected as a result. Meanwhile, the more efficient minimill industry should grow from 16 million tons (1980) to 37 million tons in the 1990s.[17] (A recent study predicts that minimills will account for 25 to 30 percent of U.S. capacity in the latter part of the 1980s.)[18]

Whether a restructured, more competitive U.S. steel industry can recapture its 1980 share of the domestic market (about 85 percent) without further government restrictions on steel imports is difficult to predict. The industry probably will be able to retain the bulk of the domestic market for most steel products, and some steel economists believe that even without tariff protection, foreigners could not capture more than one-third of the U.S. market.[19] A one-third aggregate civilian and defense-industry dependence on foreign sources would probably not endanger U.S. national security, even during an all-out war that interrupted overseas transportation routes: under such conditions, effective civilian demand for steel products would fall drastically.

Aluminum

Although the U.S. competitive position in aluminum is stronger than in steel, there are certain parallels worth exploring. In 1961 this country accounted for 49 percent of non-Communist world production of primary aluminum and was a major exporter. But by 1980 this had fallen to 34 percent; the United States even imported a small amount of its net aluminum requirements during most of the 1970s. The early 1980s saw considerable overcapacity in the world's aluminum industry as a result of both the sharp drop in long-term world growth of aluminum consumption (beginning in the early 1970s) and the 1982–83 recession. During that recession, U.S. primary aluminum producers operated at about 60 percent of capacity and some of the shut down plants are unlikely to be restored. In addition, several new aluminum smelter projects were canceled or postponed, in contrast to Australia, Canada, and South America, where expansion continues.

The principal cost factor determining competitive advantage in aluminum is electric power, and the era of relatively low-cost power in the United States is over. (Comparative power costs in various countries are discussed in app. 5-3.) Several U.S. aluminum companies are investing in new smelting capacity in countries with low-cost hydroelectric power rather than at home.

Longer-Run Outlook

Projections of U.S. dependence on aluminum imports by the end of this century vary widely, depending in part on anticipated rates of U.S. consumption growth. A recent World Bank study estimates that U.S. primary aluminum production will increase to 4.5 million mt by 1995, while consumption will rise to 5.8 million mt—leaving nearly 0.9 million mt to be filled by imports (after allowance of 0.4 million mt for secondary production).[20] A more pessimistic forecast, which assumes a higher rate of growth in U.S. aluminum consumption (5.3 percent per annum over the 1980–2000 period), foresees primary production at 4.8 million mt and consumption at 16.2 million mt in the year 2000, with 11.4 million mt to be supplied from imports. This assumes that little or no additional capacity will be built and that the United States will not use import restrictions to stimulate domestic productive expansion.[21]

The second forecast appears too pessimistic. Its estimated rate of growth in U.S. consumption (the same as that made by the BOM in 1978) is considerably higher than more recent projections. Moreover, the production figure appears to take no account of recent technological developments that reduce energy requirements for producing aluminum, or of gains to be made by expanding secondary production from scrap (a process requiring only a fraction of the electric power needed for primary production). Aluminum recovery from old scrap (e.g., cans) has been increasing rapidly in recent years; in 1981 it accounted for 15 percent of consumption.[22]

Although the United States seems likely to become dependent on imports for up to one-third of its primary aluminum needs by the end of this century, that would not threaten national security. Ample capacity for military and essential civilian requirements during a period of national emergency would remain. Moreover, aluminum metal as well as raw materials for producing it can be stockpiled for emergency use.[23]

The Social Costs of Protectionism

All import restrictions, whether in the form of tariffs, quotas, or other ntbs, mean consumers must pay higher prices for the products involved. It may be argued that the consumers' losses are offset by gains to workers and owners of protected industries, but protection entails social costs that go beyond the transfer of income from one domestic group to another. Social costs occur when domestic resources are used to produce goods available more cheaply from abroad. Moreover, imports are ultimately paid for by exports, and export industries are more efficient than industries producing for domestic markets that require import protection or some form of subsidy. Import restrictions that ultimately reduce U.S. exports sacrifice employment and output in the more efficient industries in favor of less efficient, protected industries. This amounts to a net social cost—a reduction in real national income.

The argument that the social cost of trade protection is offset by increased income from higher production and employment assumes that imports can be curtailed without reducing exports. For a large trading country such as the United States this is a false assumption. Not only will domestic protection reduce foreign purchasing power for U.S. goods or invite retaliation against U.S. exports, but in the case of nonfuel minerals, raising prices above world levels will make it harder for manufacturing industries using these materials to compete in foreign markets. Both these points are illustrated by the analysis later in this chapter of how U.S. import restrictions affect copper and steel.

Import restrictions on nonfuel minerals involve a special kind of social cost to the United States, because a large percentage of our nonferrous metal imports and a growing volume of steel products come from developing countries.[24] Low prices for nonfuel minerals have contributed to the external debt problems of mineral exporters such as Brazil, Chile, Mexico, and Peru, and U.S. import restrictions on these commodities would exacerbate their financial difficulties. Not only does the United States have a direct national interest in the welfare of friendly Third World countries, but its financial institutions have an important stake in their capacity to service their debts.

Vulnerability to Import Disruption and Maintaining the National Industrial Base

Vulnerability to import disruption refers both to particular materials, such as copper and bauxite, and to the products of an entire industry, such as steel. The distinction was made earlier in this chapter between vulnerability in peacetime and during a major war when most international commerce is disrupted, and the national and world market system is replaced by allocation and price controls. Regardless of the improbability of a long conventional war during which the U.S. production system would be seriously damaged, we must deal with the argument for import restrictions premised on this possibility.

To begin with the peacetime scenario, vulnerability to severe import disruption applies to only a handful of nonfuel minerals, the production of which is heavily concentrated where political instability is endemic or likely. Although the United States is vulnerable to import disruptions from particular sources of major metals such as copper and zinc, the impact on its economy would be limited to an abnormal rise in prices for the metals. Moreover, a high degree of self-sufficiency in, say, copper would not protect the economy from a sharp rise in world copper prices unless the United States imposed export restrictions and held down domestic prices by allocation and price controls. Without controls, domestic and world prices are closely linked.

In the case of steel products, the principal U.S. imports come from developed countries where civil disturbances or regional conflicts are unlikely, and in any case, no more than one of the several sources is likely to be disrupted at any given time. Temporary reductions in output may occur because of strikes in any of the industrial nations, including the United States, but large worldwide disruptions could take place only in the event of global war.

Turning now to the conventional-war scenario, this country should rely principally on stockpiling to assure itself of an adequate supply of minerals for which it depends heavily on imports. For reasons discussed in chapter 7, achieving domestic autarky by high-cost production is much more expensive than stockpiling. However, even stockpiling may be unnecessary for commodities like copper or iron ore for which the United States is relatively self-sufficient, perhaps as much as two-thirds. The degree of self-sufficiency necessary will depend upon estimates of requirements for military and essential civilian production in wartime. Stockpiling a large variety of manufactured goods, such as iron and steel products, is not feasible. Both civilian and military demand for individual steel products is constantly shifting, and technological developments require continual change in design and other features.

It is often argued that as a world power responsible for the defense of Western countries, the United States should have a large and diversified iron and steel industry. Some apply this argument not only to an all-out war, but use it to maintain that such an industry is necessary to support domestic enterprises that require its products. This position is justified when it comes to national security. In periods of emergency this nation may need to mobilize its industrial base for rapid production of military goods, which could be a problem if we are heavily dependent on foreign sources for a variety of steel products. But large and diversified machine-tool, computer, electronics, and other industries are also vital to defense and to essential civilian production. There is no valid reason for singling out just one industry from among the complex of interdependent domestic industries for special treatment.

America therefore faces a dilemma. On the one hand national security requires a large and well-diversified industrial base capable of meeting sudden changes in the level and pattern of demand for its products, while on the other hand U.S. industries belong to an interdependent world network of production. Autarky for all (or most) essential products *at each stage in their production* would require a maze of government controls on trade and production to restructure each industry. Import protection for one stage of production alone would not achieve independence from imports needed to make other products. Promoting a high degree of industrial self-sufficiency would devastate U.S.

trade and cause severe damage to the economic welfare and trade of our allies and the developing countries. An "economic fortress America" cannot be justified given the social costs involved and the low probability of a long conventional war.

The proper course is to keep the economy fully integrated with the international trade network, strengthening essential industries where necessary by measures other than trade restrictions. Tax incentives and other forms of subsidization would be more cost-efficient and less damaging, both to consumers and to domestic export industries, than import restrictions.[25] The steel industry, however, is not one of those with a legitimate need for special measures in the interest of national defense.

Import Protection for the U.S. Steel Industry

Since 1968, steel has been subject to a higher level of import restriction than any other category of nonfuel mineral products. In December of that year the government negotiated "voluntary restraint arrangements" (VRAs) with Japan and European countries, limiting their exports of steel products to a target of 14 million tons in 1969, after which the target would increase gradually year by year.[26] The VRAs lapsed in 1974, but the Reagan administration again made them the principal instrument for restricting steel imports.[27]

In the early 1980s U.S. steel producers filed a series of unfair trade practice complaints, mainly against EEC exporters of several categories of steel.[28] In August, 1982, the government reached an agreement with the EEC, limiting its exports of carbon and alloy steel products to an average of 5.4 percent of the U.S. market. Similar restrictions were negotiated for pipe and tube products. Altogether the agreement covered about 90 percent of European steel exports to the United States.[29] In exchange, the United States persuaded the domestic steel industry to withdraw some forty unfair trade complaints filed against European producers.[30] If this VRA had not been reached and antidumping and countervailing duties had been imposed on European steel products instead, the EEC would undoubtedly have responded with additional restrictions on U.S. agricultural and manufactured products. A number of unfair trade complaints were also filed against non-EEC countries, including Argentina, Brazil, Mexico, and South Korea. In addition, the U.S. steel industry lobbied heavily for legislation to limit carbon steel imports to 15 percent of the market for five years. Then Democratic presidential candidate Walter Mondale strongly supported this bill.

In January, 1984, Bethlehem Steel Corporation and the United Steelworkers of America jointly petitioned the International Trade Commission (ITC) for relief from serious injury under Section 201 of the Trade Act of 1974.[31] The petition asked that admission of foreign carbon and alloy steel (accounting for 90 percent of all steel mill imports) be limited to 15 percent of the U.S. market. In the July, 1984, ITC report to the president, three of the five commissioners determined that carbon and alloy steel products are entering the United States in such increased quantities as to cause serious injury to domestic producers. The three commissioners recommended a five-year program of tariffs and quotas for certain carbon and alloy steel imports, but with tariff rates and/or quantitative restrictions applied differentially to nine classifications covering most imports of these products. Two commissioners opposed tariff or quota relief.[32]

One of the dissenting commissioners, Paula Stern, pointed to other causes of reduced demand and profits in the integrated steel industry, including (*a*) the secular (long-term) decline in requirements for steel products; (*b*) high real rates of interest; (*c*) competition from U.S. minimills; (*d*) high wage rates in the industry; and (*e*) failure to employ new technology.[33] She also stated that the import restrictions advocated by the ITC's majority were "more likely to inhibit rather than enhance the overdue efforts of U.S. steel producers to adjust to conditions of competition of the 1980s."[34] Commissioner Susan Liebeler, who also dissented, cited evidence that "the import relief requested by petitioners would result in a net social welfare loss to the economy of several billion dollars. The loss to consumers which would flow from the import relief recommended by the Commission majority is also far greater than any revenue gain to the domestic steel industry."[35]

Import Restrictions and the Future of the U.S. Steel Industry

The ITC dissenting opinions advanced two fundamental arguments against import restrictions as a means of assisting the U.S. steel industry, both of them reflecting valid considerations of social cost-benefit accounting. In the first place, the social costs of import protection would outweigh their social gains. The social costs included not only the inevitable rise in domestic steel prices but the curtailment of U.S. exports due to foreign retaliation. U.S. exports of fabricated products using steel, such as factory machinery and farm equipment, would also suffer because the rise in domestic steel prices would make them less competitive in international markets. Second, protection would not cure the basic problems of the industry; it would, in fact, delay rather than promote such measures as plant modernization and cost reduction essential for an improved competitive position.[36]

On September 18, 1984, President Reagan rejected both direct quotas and tariffs in favor of negotiating VRAs with all foreign steel-exporting countries. The stated objective is to limit imports of most iron and steel products to 18.5 percent of the domestic market.[37] If the administration succeeds in this undertaking, each country or country group, such as the EEC, will be given a fixed percentage of the U.S. market, which would virtually eliminate competition by foreign suppliers; consequently both domestic and foreign producers would raise their prices. Some foreign countries would surely also impose further restrictions on U.S. exports as a condition for agreeing to VRAs.[38] Global division of the U.S. market would constitute the worst possible outcome in terms of the social costs to consumers, to U.S. exports of fabricated products, and to an efficient domestic steel industry. The proposed VRAs would also reduce export earnings of Third World countries such as Brazil, South Korea, and Mexico, all of whom are now major suppliers to the U.S. market.

As noted earlier in this chapter, steel economists do not believe that imports could take over more than one-third of the domestic market, even in the absence of import restrictions. However, imports *have* been forcing the industry to reorganize and modernize. Several efficient U.S. firms already compete well against imports. Also, new capital, including foreign investment, moved into domestic steel production in 1984.[39] One key to restructuring requires the industry itself to import more semifinished steel with which to make finished products that are competitive in both domestic and foreign markets. Thus internationalization can actually contribute to the strength of domestic industry.

Import Protection for Nonferrous Metals Industries

The principal nonferrous metals that face strong foreign competition in U.S. markets are copper and zinc. Both commodities have been the subject of petitions to the ITC for import relief under the Trade Act of 1974. Although these petitions were denied in 1978, the industry made another effort to obtain a copper import quota in 1984. In addition, bills have been introduced in Congress to protect the copper industry from foreign competition.[40]

Unlike steel, there is no valid argument for linking any particular degree of domestic self-sufficiency in copper or zinc to national security. Foreign supply sources are broadly diversified, with a substantial percentage of imports coming from Canada and Mexico. There is no significant vulnerability to supply disruption in peacetime. For the contingency of a long conventional war, potential vulnerability can be avoided at a far lower social cost by stockpiling than by protection from imports.

The arguments against protection for nonferrous metals are well illustrated by an ITC finding against major U.S. copper producers seeking an import quota. In the ITC report of July, 1984, all five commissioners found that increased copper importation is a substantial cause (or threat) of serious injury to the domestic industry, but a majority of the commissioners opposed quantitative import quotas. This position was upheld by President Reagan in

September, 1984.[41] One commissioner pointed to a number of factors that were more important sources of injury than increased imports. These included the relatively high costs paid by domestic producers for labor, environmental protection, and energy compared to their competitors abroad; and a short-term cyclical decline in demand for copper combined with a long-term downward trend in the rate of growth of demand.[42] The commissioners argued that a quota on copper metal would not help the copper producing industry but would harm the industries producing products made from copper. This is because a quota would raise the domestic price of copper metal above the world price, with the consequence that domestic producers of copper products would be unable to compete with foreign producers and there would be a flood of imports of copper products. This would reduce domestic demand for copper metal as well as for domestically produced copper products.[43]

The same argument applies to import restrictions on other nonfuel minerals. By raising the domestic price of a metal well above the world price, domestic fabricators using the metal are unable to compete with fabricators abroad in either domestic or foreign markets. This would reduce demand for domestically produced metal and thereby offset the effects the import quotas were intended to bring about.

Conclusions

There is no necessary relationship between domestic output of most nonfuel minerals and their availability for civilian and defense requirements. For those of which the United States is a substantial producer, foreign sources are well diversified: no governmental measures are necessary to protect the economy against import disruption from a particular country or region during peacetime. In the unlikely event of a long conventional war that denied this country all import sources except Canada and Mexico, supplies of nonmanufactured metals such as copper and zinc could be assured by stockpiling. Import protection is not only unnecessary to insure domestic availability of nonfuel minerals, but tends to delay plant modernization, cost reduction, and adoption of advanced technology. Moreover, the social costs of protection exceed any conceivable social benefits. These social costs include price increases for consumers, reduced competitiveness in world markets for industries forced to utilize artificially high-priced minerals in fabrication, undue hardship for Third World mineral-exporting countries struggling to service their external debt (much of it to U.S. banks), and foreign retaliation against U.S. exports.

The worldwide competitive position of the U.S. steel industry has been declining since at least 1960, but, as pointed out earlier in this chapter, many in the industry continue to thrive, and most of the industry can survive competition from imports. Protection would not encourage modernization; it is more likely to prolong the use of old equipment and delay the building of more efficient plants. In the longer run, developing countries such as Brazil and South Korea will supply an increasing share of U.S. and European markets for steel. The developed countries best serve their own economic interests by opening their markets so that developing countries can not only sell them less expensive materials, but thereby earn the foreign exchange they need to buy agricultural products and sophisticated manufactures from their trade partners.

If protective measures are not the way to promote the viability of the U.S. steel industry, what alternatives are available? Some have suggested various forms of subsidies, such as liberalized tax regulations to promote new investment. Although subsidies are superior to protection, they will not guarantee performance or correct fundamental internal problems. The United States does not face an imminent collapse of the steel industry, and the industry's earnings improved in 1984.[44] Independent steel economists maintain that the best therapy is for competitive forces to bring wages down to levels comparable with other manufacturing industries and to carry through a financial and managerial reorganization that will lead to increased efficiency. The performance of some firms in the industry and the willingness of foreign companies to invest in it belie the argument that U.S. steel products cannot compete effectively in world markets.

There is no economic or national security basis for imposing a burden on American taxpayers and consumers to promote the special interests of the mining and metal-processing industries. Their representatives lobby tenaciously against government regulations that interfere with their freedom of operation. It is ironic that these same industries are asking the government to impose trade restrictions on other U.S. producers, including other segments of the metal industry that need imported products. The metal industries thus champion principles of free competitive enterprise on the one hand, but on the other they call for government support that would disadvantage and interfere with the freedom of others to buy in the cheapest markets. Behind the cloak of national defense they seek to promote their special interests at the expense of the international competitive strength of industries using their products.

As noted in chapter 3, several large U.S. metal companies have foreign affiliates making the same products abroad that the parent company does at home. For example, ASARCO and Newmont, two of the largest domestic copper firms, produce more abroad than they do in the United States. In principle, a multinational corporation ought to support free trade in all countries. But ASARCO and Newmont support import quotas on copper, and ASARCO officials are strong advocates of import restrictions on other nonferrous metals as well. A number of steel concerns have affiliates in Canada from whom they import steel products they use in the U.S. mills. These firms nevertheless advocate steel import quotas. Apparently these multinationals believe they can gain more from import restrictions in support of their domestic operations than such restrictions would cost their foreign operations. I believe, however, that after taking into account retaliation by foreign governments and other adverse impacts on world trade, U.S. import restrictions would reduce the global profits of U.S. multinationals.

Appendix 5-1

Cost Comparisons in the Copper Industry

Cost comparisons between the mining industries of different countries are hard to make, partly because there is little comparable data, partly because mines in the same country show large cost differences—differences that have little to do with direct production costs such as labor and energy. In addition to varying ore grades, costs are greatly affected by ore by-products, such as gold, silver, and molybdenum. (By-product values are deducted from total costs to arrive at production costs of the principal mineral.)

The average grade of copper ore mined in the United States during the 1971–75 period was estimated at 0.52 percent, as contrasted with 1.6 percent during the 1930s. This compares with an average grade of 1.03 percent in Latin America and of 1.67 percent in Africa.[45] However, as just noted, the average grade does not necessarily correspond to the net cost of extraction. For example, it might be profitable to mine an ore body with only 0.4 percent copper if it contained a substantial amount of gold as a by-product. (This is why the Bougainville copper mine in PNG has been so profitable.)

Most cost estimates are made for individual mines in various countries. Cost comparisons between countries require a special method to define certain costs. For example, interest accounts for a large part of the cost for a mine that is heavily financed by debt, but not for one that is mainly financed by equity or whose debt has been repaid. Typically, new foreign mines are heavily financed with debt. Royalties qualify as costs, but in some countries they are low or nonexistent (as in the United States), and the government extracts revenue by taxes on net profits—which are not regarded as a cost. Depreciation is a cost item, but the rate of depreciation permitted varies from country to country. A proper comparison should be based on full economic costs, including a return to risk capital sufficient to attract new investment, but data for this purpose are seldom available.

Cost comparisons between countries require selection of an exchange rate to convert domestic currency into a common one, usually the U.S. dollar. For many developing countries, including Chile and Zambia, the official exchange rate has been highly overvalued in recent years, thereby artificially *raising* apparent domestic costs when converted into dollars. Some Chilean copper cost estimates are much higher than others, depending upon the exchange rate used when the calculations were made. The dollar itself appreciated over 30 percent in relation to other major currencies between 1980 and early 1984. This had the effect of raising U.S. production costs relative to those of other countries when their currencies were converted into dollars. Hence 1982 or 1983 comparisons of U.S. with foreign production costs are distorted by a high exchange value for the dollar.

One cost study provides a comprehensive comparative analysis of major mines and countries for 1978. Although all mining costs (in constant dollars) have risen since then, if all the effects of higher U.S. pollution abatement costs and appreciation of the dollar are excluded, U.S. costs probably did not change much between 1978 and 1984 relative to those in most other major copper-producing countries. According to this study, the weighted average cost for twenty-eight U.S. copper mines was $0.71 per pound, as contrasted with the weighted average cost for mines in all countries in the survey (covering most of the non-Communist world producers) of $0.59. These costs refer to production of refined copper and include mining, milling, administration, transportation, smelting, refining, and local mining taxes and/or royalties. Depreciation is included, but not profit-based taxes or interest charges on debt.[46]

According to these estimates, average production costs in Australia ($0.55 per pound), Canada ($0.42), Chile ($0.46), Papua New Guinea ($0.43), and Peru ($0.59) were considerably lower than those in the United States, while average costs in Zambia ($0.74) and Zaire ($0.80) exceeded their U.S. counterparts (see app. table A5-1). It should be noted, however, that costs for individual U.S. mines in 1978 ranged from $0.55 to $1.00 per pound. In a low-cost copper producing country such as Canada, the range was $0.40 to $0.77; and in Chile it was from $0.32 to $0.64. Thus, costs in some U.S. mines were lower than those in some countries whose weighted average was substantially lower than the U.S. figure.[47]

It should be emphasized that these estimates include neither full returns

TABLE A5-1. Weighted Average Production Costs per Pound of Copper for Leading Copper-Producing Countries, 1978, in U.S. Cents per Pound

Country	Cost	Country	Cost
United States	71	Philippines	52
Chile	46	Australia	55
Canada	42	South Africa	48
Zambia	74	Papua New Guinea	43
Zaire	80	Indonesia	60
Peru	59		

Source: C. G. Streets, "The Cost of Primary Copper in December 1978," mimeographed (London: Consolidated Gold Fields, July 1979), p. 5.

Note: Costs include depreciation, but exclude profit-based taxes and all returns to capital.

on capital investment nor taxes on net profits. Profit taxes in Peru are currently so high that private investment in the copper industry there is less attractive than in the United States, although average costs in Peru are considerably lower. Another important difficulty with cost comparisons as a measure of relative competitive advantage is that most data apply to old mines: they do not reflect the full economic costs for new mines or expansion of existing mines that prices would have to cover before investment in new capacity would be attractive. Another problem is that state-operated mines are often taxed less than private mines, and that state mining concerns usually obtain capital at a lower cost than private enterprises. Also, state enterprises are often subsidized in other ways.

In order to maintain employment, state enterprises sometimes operate even when revenues are less than operating costs. Private mining companies in the United States and Canada usually cut back production corresponding to reduced demand so that producer prices can be maintained or selling below cost avoided. Given the average 1978 U.S. production costs of $0.71 per pound, and adjusting this by the rise in the wholesale price index, average costs were $1.01 per pound (excluding profit or interest on investment) by the end of 1982. Since prices were then only about $0.78 per pound, the severe U.S. copper production cutback of 1982 is not surprising.

Appendix 5-2

Structural Factors in the Steel Industry

A number of structural factors help explain why the U.S. competitive position in steel has declined. Some reflect long-term changes in world trade patterns, while others are peculiar to the domestic industry itself. Perhaps the most significant factor is the slow growth rate of U.S. consumption, an average 1 percent annually between 1950 and 1981. Among the industrial countries, only the United Kingdom's rate of growth was lower. By contrast, over the same period steel consumption in Japan and the major continental West European countries (Benelux, Germany, Italy, and France) grew at respective averages of 9.8 percent and 3.6 percent yearly.[48] These differences led to parallel differences in rates of production growth between the United States and other major industrial producers. This is one reason for the respective competitors' disparity in modernization rates and adoption of new technology. A related cause of such disparity was the emergence of new industrial countries like Canada and Australia and of newly industrializing developing countries (NICs) like Brazil and South Korea as important steel producers.

A second factor has been the worldwide excess capacity resulting from steel consumption's virtual stagnation during the late 1970s and early 1980s. From 1949 to 1969 world steel consumption grew at about 6.6 percent per year, but in the 1969–79 period the annual rate fell to 2.7 percent, with a further decline in 1980–82.[49] During 1969 to 1981 average annual consumption growth was negative for the United States, the United Kingdom, and continental Western Europe, and was positive by only 1.3 percent in Japan.[50] This structural decline in the growth of steel consumption was caused by lower general rates of economic growth combined with lower steel use intensity (i.e., the drop in the ratio of steel consumption to GNP). The latter reflected a shift to new and lighter materials in the production of goods such as automobiles, increased substitution of aluminum and lighter-weight alloys for steel-intensive products, and the increase of services as a component of GNP.

World steel-making ability grew much more rapidly than steel consumption, particularly in the EEC and Japan. Over the 1960–78 period EEC capacity rose annually by an average 4.6 percent; that of Japan grew at an annual rate of 9.7 percent.[51] Latin American steel output rose fivefold during this time, Asian output (excluding Japan) nearly sixfold.[52] (The corresponding U.S. figure was 0.3 percent.) World overcapacity stimulated Japanese and

European producers to compete more vigorously in foreign markets. The United States was their principal target because domestic steel prices were above world prices. U.S. costs were substantially above Japanese costs, but not above EEC costs. The integrated steel industry was also at a disadvantage compared to the minimill steel producers.[53]

A third structural factor making for relative decline in U.S. ability to compete was the depletion of high-grade domestic iron ore deposits simultaneously with the development of large, rich iron ore deposits in Australia, Brazil, Canada, and elsewhere. This, plus generally lower transportation costs, eliminated the earlier U.S. advantage in raw materials.

The U.S. competitive position in world steel has also been affected by specifically domestic industrial factors. The first concerns the U.S. steel industry's market structure. Unlike the copper industry, which is relatively competitive and where producer prices have stayed relatively close to world prices in recent years, for most of the period 1960–82 U.S. producer prices for steel were substantially above Japanese and European prices (20 and 30 percent, respectively, in 1981).[54]

A second factor impairing the U.S. competitive position is high labor costs due to the strong bargaining position of the United Steel Workers, together with the industry's acquiescence in wages much higher than those in other U.S. manufacturing industries. In 1982 the average hourly wage (including fringe benefits) in the U.S. steel industry was $23 as contrasted with $10 in Japan, $12 in West Germany, and $8 in the United Kingdom.[55] In 1980 U.S. wages in steel were 64 percent above those in manufacturing, but productivity (in terms of value added per worker) had not kept pace with rising wage rates during the 1970s. Labor productivity (measured by physical output per man-year) increased only 27 percent between 1970 and 1980, compared with 85 percent for Japan. At the end of that period, absolute labor productivity in the two countries was about the same,[56] and the average production cost advantage held by Japan over the United States was about $50 per ton,[57] or roughly 15 percent of the 1980 composite selling price.

A third domestic factor undermining the U.S. competitive position has been the relatively low percentage of total capacity employing new technology, particularly in such areas as continuous casting, direct reduction, and production-control equipment.[58] This reflects in part slow capacity growth relative to that in Japan and the EEC, and in part a reluctance to close old facilities. Only two new integrated steel plants were built in the United States after World War II, the last one in 1967. Investment has been inadequate, largely because low and erratic earnings make it hard to attract equity and loan capital. For the 1967–78 period, the return on equity investment after taxes was 7.1 percent as contrasted with 14.4 percent for all U.S. manufacturing.[59]

Profits in the industry between 1960 and 1980 were substantially lower every year, except 1974, than the U.S. manufacturing average.

A final aspect of the U.S. steel industry that hinders its competitiveness concerns government restrictions. Throughout much of the 1960s and 1970s, U.S. steel companies were kept from raising prices in periods of high demand and inflation, either by formal price controls or by government suasion. In addition, legal constraints ruled out some cost-saving mergers and joint ventures.

European governments, on the contrary, have been heavily involved in their steel industries, both as owners of nationalized steel firms (in the United Kingdom, France, Italy, and Belgium) and as providers of inexpensive loans and subsidies. Government grants and concessionary credits have financed steel modernization in the EEC, and the ECSC has encouraged steel firms to coordinate production among themselves. Although environmental standards comparable to those in the United States apply to European and Japanese concerns, the regulations have been applied with greater consideration of their economic impact on the industry. (In some cases government has itself borne part of the costs.) Beginning with President Truman's threat to nationalize the industry in 1952, U.S. government attitudes have been characterized by confrontation rather than cooperation. U.S. assistance to the domestic steel industry has mainly taken the form of various types of import protection, but even in this area policies have been erratic. Although U.S. trade policy has reduced foreign competition in recent years, this will probably be counterproductive for the long-run competitive position of the industry.

Appendix 5-3

Electric Power Rates and Location of New Aluminum Smelting Capacity

The availability of low-cost electric power was a key factor in the rapid growth of aluminum capacity in the United States prior to the 1970s. Access to cheap hydroelectric power from Niagara Falls, the Tennessee Valley Authority (TVA), the Bonneville Power Administration (BPA), and other sources in the Pacific Northwest led to a concentration of aluminum smelters in these areas. Low-cost natural gas for power generation also promoted the establishment of some plants in Texas, Louisiana, and other southern states. However, the era of low-cost power came to an end in the late 1970s.

In 1976 the BPA supplied electricity at rates as low as 2 mils per kilowatt hour (kwh) but the average cost in the Pacific Northwest was about 9 mils per kwh in 1979. By October, 1982, the rate had risen fivefold, and an additional 9 percent increase for industrial consumers was announced in October, 1983.[60] Nevertheless, BPA prices are somewhat lower than the average for nonhydropower systems. TVA rates have also been increasing; in 1982 they were 32 to 35 mils per kwh.[61]

A 1981 report prepared for the TVA estimated the total cost per pound of aluminum at $0.64 (in 1981 U.S. dollars) as contrasted with $0.53 in Australia; $0.51 in Brazil; $0.60 in Canada; and $0.51 in the Middle East.[62] Alumina as well as power costs less in Australia and Brazil than in the United States. The report stated that the introduction of titanium diboride cathodes in existing U.S. smelters during the 1990s is likely to produce an effective capacity expansion of approximately 10 percent, and that cost differentials between U.S. production and that of foreign countries will probably lessen.[63] The report concluded that, as a consequence, half the additional capacity for increased U.S. demand during the 1990s might be domestically located.

Because of rising electric power costs and the difficulty of obtaining long-term contracts for power at specified prices, some projects for aluminum smelter expansion have been canceled or postponed. For example, in January, 1983, Alumax Pacific Corporation announced the indefinite postponement of its planned 220,000 ton per year plant at Umatilla, Oregon.[64] Spokesmen for several large U.S. firms have stated they do not expect any new aluminum smelters to be built in this country for years to come, and that their companies expect instead to invest in smelting capacity abroad, particularly in Canada,

Australia, and South America. A number of projects have been announced. For example, Alcan is planning a 500,000 mt per year smelter in British Columbia, and Anaconda a 300,000 mt per year smelter in Labrador in joint venture with the Newfoundland government.[65] An Alcoa/Shell/Brazil refinery-smelter complex for producing 100,000 mt of aluminum per year is planned for Sao Luis, Brazil; Reynolds, in joint venture with Venezuelan companies, plans a 320,000 mt per year aluminum smelter at Alcasa, Venezuela. Reynolds is also planning to expand its facilities at Baie Comeau, Quebec, from their current 175,000 mt capacity to 300,000 mt, with hydropower to be provided by a Quebec government power company. Alcan, Alcoa, and Reynolds intend to build a number of smelters in Australia, including one at Portland, Victoria, and another in Queensland.[66] If all these are actually completed, it would add over a million mt to Australia's capacity. In addition, non-U.S. companies plan to construct additional Australian capacity of over a million mt.[67] Some of the Australian projects have been deferred, however, because of either low world demand or difficulties in obtaining power contracts. Some Brazilian projects have reportedly been delayed as well. Nevertheless, most new aluminum smelter and alumina refinery capacity will probably be located in countries with abundant bauxite reserves and hydroelectric power. Some smelter capacity will be built in Middle Eastern countries with abundant low-cost natural gas.

6

International Issues Relating to Nonfuel Minerals

The preceding chapter examined the case for restrictions on nonfuel mineral imports to advance domestic prosperity or national security. Thorough scrutiny reveals the arguments for protection to be misguided at best, specious or hypocritical at worst. In any event, social cost-benefit analysis demonstrates that both economic and defense interests are best served by continued adherence to traditional free trade policies.

Can the argument against insularity be taken beyond open markets to embrace even more comprehensive reliance on international cooperation? After all, the advanced industrial countries have a surfeit of the capital and know-how Third World nations currently require to develop the exports they depend on to finance rapid economic progress. And what many of the LDCs have is mineral resources of incomparably higher yield than those of most of their industrialized trading partners, who stand to profit directly and indirectly from the development of new production capacity abroad. Each party has something the other needs—a situation definitive of a positive-sum game. The missing catalyst is the kind of trust required for binding multilateral development agreements administered by impartial multinational agencies. The prognosis for this kind of international symbiosis is not as favorable as it should be, largely because of fundamental conflicts over the proper approach to development and international economic agreements themselves.

Prices of minerals fluctuate more widely than those of most other commodities, and price stabilization at levels consistent with profitable operation is of major concern to the welfare of mineral-exporting countries. In practice, spokesmen for developing countries often use the term "stabilization" loosely to mean support for levels they deem "fair," instead of in the more traditional sense of reducing the amplitude of fluctuations above and below long-term equilibrium price. The U.S. government has been sympathetic to international commodity agreements aimed at stabilization in the latter, not the former, sense.

In general, Third World countries have favored obtaining external capital for their mineral projects from international assistance agencies such as the World Bank rather than from multinational mining firms. The governments of the United States and other industrial countries, however, have maintained that foreign financing for these projects should be left to the private sector.

The discovery of rich seabed mineral nodules has raised the question of ownership and control of these resources, which are regarded by LDCs as the common property of all nations rather than that of private companies finding and exploiting them. This issue has been an important source of contention at the Law of the Sea (LOS) Conferences and in the formulation of the Law of the Sea Treaty.

Two more international issues are of special concern to industrial countries. One is unfair international trade practices and import restrictions on steel products, an issue also of special interest to a few nonindustrial countries that export steel. The other concerns international cooperation for dealing with mineral supply disruptions due to civil disorder or limited war.

International Commodity Agreements

Chapter 4 showed that both cyclical and long-term nonfuel minerals price fluctuations have caused mineral-producing capacity to develop out of step with the long-run growth in world demand. Price fluctuations that bear no relationship to changes in the cost of production are burdensome to producers and consumers alike. These irrational variations have been particularly hard on Third World nonfuel mineral-exporting countries; consequently, they have sought to moderate price swings through international commodity agreements. Developed countries that are important producers of nonfuel minerals, such as the United States and Canada, are adversely affected by the same price oscillations, while consumers of the minerals everywhere would benefit if the magnitude of these fluctuations were reduced. And yet the outlook for international cooperation in price stabilization is not promising, because the approaches and objectives of developing countries have differed sharply from those of the developed countries in recent years.

The Third World effort to negotiate international agreements to control the prices of primary commodities has found expression in the Integrated Program for Commodities (IPC), strongly endorsed by developing countries at the Fourth United Nations Conference on Trade and Development (UNCTAD) in Nairobi in 1976. The IPC calls for international accords on ten core commodities, including the nonfuel minerals copper and tin. Other price stabilization measures would affect seven additional commodities, including bauxite and iron ore.[1] The agreements would include both producing and consuming countries. In addition, the IPC provides for governments to establish a common fund of $6 billion. When the IPC was formulated, four of the ten core commodities—cocoa, coffee, sugar, and tin—were already covered by formal accords. (From time to time, the United States has been a party to the coffee and tin agreements.)[2]

The United States and other major developed countries gave their general if unenthusiastic approval to the IPC in 1976, but there are substantial differences between the developed countries and the UNCTAD secretariat (whose policies are dominated by Third World countries) over its implementation. For the developed nations, price stabilization involves moderating price variations above and below a long-term trend so that they approximate the long-run equilibrium price. Such a goal can best be achieved by a buffer stock arrangement in which a supply of both money and the commodity make up the stock. The buffer stock manager would buy the commodity in periods when the world price fell below the estimated long-run equilibrium price by a certain percentage and sell it when the price rose above equilibrium by a certain percentage. The success of such an operation would be indicated by the reestablishment of the initial composition of the buffer stock (in money and commodity) from time to time over a period of, say, three to five years.

The UNCTAD secretariat and Third World countries, by contrast, would impose export restraints (quotas) to supplement buffer stock operations whenever there are insufficient funds to maintain an international price regarded as "fair" (often equal to some historical level) for the commodity. The two approaches epitomize the difference between true commodity stabilization and the setting of a floor price. A buffer stock arrangement does not interfere with freedom of the market, but operates within its context by influencing demand and supply. Export quotas, on the other hand, directly restrict market operations; they might maintain a floor price for the short run, but in the long run they are likely to burden the market with large surpluses.

The United States never favored creation of the common fund, and even though the Carter administration was willing to go along with it under certain conditions, it is doubtful that Congress would have appropriated money for it. Negotiations broke down in 1978 because of differences between developed countries and the UNCTAD secretariat over the fund's objectives and functions.[3] Although its financing would come from both mineral-producing and -consuming countries, such an international buffer stock agreement would get its major contributions from developed consumer countries. Hence it is not surprising that when, in June, 1980, UNCTAD proposed establishment of a common fund along these lines, the would-be large donors proved less than enthusiastic. Five years later, the fund agreement had not been ratified by the countries scheduled to provide two-thirds of its capital. Moreover, the Reagan administration is unlikely to approve a U.S. contribution.[4]

Feasibility of Commodity Price Stabilization

Commodity specialists and economists disagree sharply over whether international commodity agreements can work.[5] Economists point out that it is extremely difficult to establish the long-run equilibrium figure that would serve as the reference price, or target, for stabilization. Moreover, in practice the targets are set by negotiation between political representatives of producing and consuming nations rather than by the independent judgment of commodity specialists. Simulation studies for copper and other commodities have shown that to achieve reasonable price stability, buffer stocks would require a great deal of money, running to several billion dollars, plus a large supply of the commodity, and that the United States and other developed countries are unlikely to contribute resources of this size.[6]

Since 1976, more than a dozen meetings have been held under the auspices of UNCTAD in an attempt to reach an international buffer stock agreement in copper. Quite a number of technical studies have been conducted on the feasibility and cost of a copper buffer stock, but they reach different conclusions.[7] The latest estimate is that an effective copper buffer stock would cost about $2 billion, but given depressed prices that are below costs for most mines, together with uncertainties regarding the future rate of growth in demand, agreement on a target around which to stabilize copper prices would be exceedingly difficult. It seems unlikely that a copper agreement will be negotiated.

The First International Tin Agreement, which included both producing and consuming nations, was ratified in 1956 and was subsequently renewed. The record of the ITA in stabilizing tin prices has been poor, and it has been more successful in defending a floor than a ceiling price. The International Tin Council has imposed export controls in addition to relying on the buffer stock for maintaining prices above the floor. Strong differences regarding the floor and ceiling prices have hampered implementation of the agreement. The United States joined the Fifth ITA in 1976 (primarily to show its willingness to cooperate with the tin-producing countries), but it has objected to export controls and to the level of support prices, which it believed to be higher than the long-run equilibrium figure justified.[8] The Reagan administration refused to join the Sixth ITA initiated in 1982. In October, 1985, the tin buffer stock ran out of cash after having borrowed several hundred million dollars against tin it had purchased to maintain the floor price. The market price plunged well under the floor price, and the LME suspended trading in tin. In early 1986, the New York market price was less than half the floor price, and it now appears that the International Tin Council will no longer be able to exercise control over the tin market. The principal reason for the breakdown in the stabilization program is that the ITC set the price too high, with the result that producers in countries not members of the ITA expanded their output. In addition, consumers turned to substitutes for tin, such as aluminum.

Prospects for Stabilizing Nonfuel Mineral Prices

A buffer stock manager with complete flexibility for adjusting targets and ample minerals and financial resources should be able to moderate cyclical and other short-term price movements. Since international cooperation in this field involves reconciling divergent national interests rather than a rational approach to a commonly agreed objective, these ideal conditions are unlikely to prevail. And even if shorter-term fluctuations could be moderated, little could be done to avoid long-term downtrends in real prices such as have occurred since 1975, which are a response to lower secular rates of growth in demand for nonfuel minerals.

International trade and production controls for stabilizing prices are an alternative to buffer stock operations. The former were employed more or less successfully by OPEC for nearly a decade, but only the dominant position of Saudi Arabia and the willingness of a few other producers to limit petroleum production made it possible to keep export prices high. Such conditions do not exist for any major nonfuel mineral. Moreover, strong cartels in nonfuel minerals are unlikely to promote either the welfare of consumers or the discovery and development of minerals to meet long-term world requirements. Fortunately, it is politically impossible to create an international agency with power to regulate the creation of additional capacity in all producing countries and to determine how much they should export and at what prices. In short, there is simply no satisfactory alternative to the operation of world competitive forces in the nonfuel mineral industries. To paraphrase Winston Churchill's aphorism about democracy, international competition is the worst possible arrangement, save for all the others!

International Assistance to Mineral Development in the LDCs

International support for the mineral industries of developing countries has included technical assistance from the United Nations, loans from international agencies such as the World Bank, and bilateral assistance from governments. All three kinds of help have benefited state mining enterprises, private domestic firms, and foreign private investment (the latter usually through joint ventures between foreign firms and domestic private or state mining concerns).

Both developed and developing countries stand to gain from an increased flow of capital and technology to LDCs that want to expand their mineral industries. However, the form and direction that public international assistance should take has been controversial. The United States and most other developed countries have favored aid to private enterprise, both domestic and foreign. Not only is private enterprise regarded as more efficient than public enterprise, but it is believed that state enterprises operating in a competitive world market should not be provided with public international loans at rates that reduce their capital costs below those incurred by private foreign investors. Moreover, since international assistance is in short supply, it should not be given to an industry for which private capital is available.

The major interest of developing countries, and of the UN agencies they tend to dominate, has been to establish greater state control over natural resources. Thus they favor international assistance to public mining and processing enterprises. The principle of "full permanent sovereignty of every state over its natural resources, including the right of nationalization or transfer of ownership to its nationals," has been an important element of the New International Economic Order (NIEO) advocated by many LDCs. A 1974 UN resolution provides a program of action: efforts should be made "(*a*) to defeat attempts to prevent the free and effective exercise of the rights of every state to full and permanent sovereignty over its natural resources; and (*b*) to insure that competent agencies of the UN system meet requests for assistance from developing countries in connection with the operation of nationalized means of production."[9] Similar UN resolutions go back to the early 1950s.

The wave of expropriations in both mining and petroleum during the late 1960s and early 1970s greatly reduced the flow of private equity capital to Third World mineral industries, and there was increasing concern in both the UN and the international development agencies that neither the capital nor the technology needed for exploration and development of LDC mineral-producing capacity would be forthcoming.[10]

In 1972 a World Bank staff report assessed the growing capital requirements of nonfuel mineral industries and recommended a greatly expanded program of loans by the World Bank Group (WBG) for mineral projects, including exploration.[11] Until then the group had regarded mineral development as largely the province of multinational companies. The U.S. position favored expanded international development financing, but held that the role of the WBG and other international development loan agencies should be to act as catalysts in attracting private international equity and loans. The United States thought the International Finance Corporation (IFC) could bolster the confidence of private investors in Third World countries by taking minority equity positions and by making loans to private domestic and foreign firms. In addition, it wanted the World Bank to promote foreign direct investment through loans to joint ventures involving state mining enterprises and foreign private investors. That the U.S. position has not entirely prevailed is indicated by a number of World Bank and Inter-American Development Bank (IADB) loans to state mining enterprises. During the Reagan administration, the U.S. member on the panel of World Bank executive directors has voted against a number of such loans, but the United States does not have a veto there.

WBG and IADB Financial Assistance

Prior to 1977 the WBG made relatively few loans to the mining sectors of developing countries. Altogether, these sectors had received less than 2 percent of the total value of World Bank and International Development Association (IDA) loans. Beginning in July, 1977, the World Bank's executive directors made it their policy to expand the WBG's promotion of nonfuel mineral resource development in LDCs. The WBG was to act as an "active catalyst" for the flow of equity and loan capital.[12] The bank estimated that between 1976 and 1985 developing countries might need as much as $95 billion invested in their nonfuel mineral sectors. The bulk of these funds would have to come from private sources. The WBG was expected to lend about $1 billion over the five-year period 1978–82, allowing four to six mineral (including coal) projects per year to be financed.

Actual loans have been considerably short of the announced goal. During the July, 1978, through June, 1983, period, World Bank and IDA loans for mining totaled only $558 million, mainly to state mining enterprises.[13] For the calendar years 1978 through 1983, the IFC made loans and equity investments in private metal mining and processing projects totaling about $287 million.[14]

The IADB has made relatively few loans for mining projects in Latin America, most of them to projects wholly owned and operated by host governments. For example, it made several loans to the Peruvian state mining enterprise Centromin, a complex formerly owned by American investors under the name Cerro de Pasco before its expropriation in 1974. A 1983 IADB loan of $268 million to the Chilean government's CODELCO has been highly controversial because of the world's excess copper-producing capacity. The U.S. executive director at the IADB voted against the loan.

Two UN agencies have provided technical assistance to developing countries for their mineral industries, namely the United Nations Development Program (UNDP), which has been operating for a quarter century, and the UN Revolving Fund for Natural Resources Exploration, established in 1975. These bodies have conducted preliminary exploration for nonfuel minerals, including copper in Chile, Mexico, Malaysia, and Panama, and tin in Indonesia. Some deposits explored by these agencies have been developed into producing mines, either by private domestic or foreign investors, or by government enterprises. The UN Centre for Natural Resources and Transport and the UN Centre for Transnational Corporations have provided advisory services to Third World governments in negotiating mining agreements with multinational companies.

Opposition to Aid for State Mining Enterprises

U.S. mining interests, including the American Mining Congress, have strongly opposed World Bank, IADB, and other international public assistance to government mining enterprises. They argue that such aid subsidizes mineral production and creates excess capacity.[15] This is a problem, however, that has to be considered from the standpoint of social benefit-cost accounting rather than from that of private mining interests.

The United States and other industrial countries have an economic and political interest in promoting economic development in the LDCs through international assistance, including loans. Some argue that such assistance to government mining enterprises constitutes a misallocation of resources as contrasted with other types of development, such as agricultural productivity, education, and health. But the validity of this argument depends on circumstances in the individual borrowing country and should be determined by the development assistance agency.

Another argument is that since private international capital is available to finance Third World mining activities, provided a favorable investment climate exists, international development agencies should restrict themselves to financing projects for which private capital is not available. This contention accords provisions of the World Bank charter,[16] but Third World countries hold that such a policy would violate another provision of the charter that forbids the bank to discriminate on the basis of national political and economic structure. Hence if a country's political orientation favors government enterprise for development of its mineral resources, the bank should respect this. The World Bank has yet to resolve this apparent contradiction in its charter.

World Bank Group and other international public agency assistance for mineral development should also be considered from the standpoint of world mineral consumption. International public assistance for geological studies and exploration for new reserves in developing countries clearly serves the long-run interests of all consumers of nonfuel minerals. This kind of subsidization could be justified on the grounds that because political and economic risk in many of these countries is quite high, the optimum expenditure for exploration and development is otherwise unlikely.

Of course, public agency assistance to state enterprises for mining known reserves might not contribute to the long-run availability of mineral supplies if it resulted in overcapacity. Such overcapacity could lead to premature closure of private mines excluded from low-interest-rate loans and other forms of subsidy, which would discourage new investment. Moreover, subsidizing excess capacity constitutes a misallocation of global resources. These considerations must be taken into account by development agencies in making individual loans, but they do not constitute an argument against all public agency assistance to government mining enterprises.

The objection of private mining interests to subsidies for state enterprises raises the broader question of whether international development agencies should deny assistance to all LDC industries that compete with private firms in world markets. If so, the agencies would find it impossible to promote broadly based economic development. For example, international assistance has directly or indirectly financed steel and other manufacturing industries that compete in the markets of the United States and other industrialized countries.[17] Moreover, economic development requires expanded exports, and some of these will inevitably compete with products from the countries that provide assistance.

These issues cannot be adequately discussed here, but they do raise doubts as to whether it is legitimate for U.S. mining companies to pressure the government to vote against international agency loans to their competitors in developing countries.[18] The government should, of course, oppose loans that would misallocate development resources, but not to accommodate special interests in this country.

Tentative Conclusions on International Public Assistance

International public assistance for nonfuel mineral industries in the LDCs involves ideological and political controversies as well as difficult analytical questions regarding the allocation of resources for optimum social benefit. The World Bank Group and other international public agencies cannot provide more than a fraction of the investment capital needed by Third World mining and mineral-processing industries. Therefore, loans from these sources should be leveraged as much as possible: they should not take the place of what private foreign and domestic sources could provide. Moreover, since private international capital is available for equity and loan financing of LDC mining projects, assistance funds should not be provided, as some Third World countries argue, simply because they choose to develop their resources through state enterprises to the exclusion of foreign direct investment.

Three types of international public assistance to Third World mineral development are clearly justified. First, geological surveys and high-risk exploration for new reserves in poor countries should be promoted by technical assistance and low-cost loans when the risks involved are too great for international mining companies. Mineral discoveries could then be developed by international private capital, management, and technology. Second, modest loans and equity investments by the IFC can encourage the investment of private foreign and domestic capital in mineral development: very often the participation of the IFC in a project gives equity and loan investors confidence to provide the much larger funds needed. Third, development assistance agencies can make loans to joint ventures between private or government enterprises and foreign mining companies if such loans would promote the availability of additional foreign equity and technical and managerial skills. All three types of assistance would serve to increase the flow of private external resources to LDC mining industries.

Seabed Minerals and the UN Law of the Sea Convention

The chasm between the ideology and perceived interests of Third World and developed countries, so pronounced in the debate over commodity agreements and international public agency assistance for mining, becomes more like a complex network of fissures when seabed exploration and exploitation are concerned. For one thing, the developed nations are far from agreement on whether to endorse the proposed Law of the Sea. For another, even in the United States political leaders are divided over the relative merits of signing or abstaining. In any case, no assessment of the future availability of nonfuel minerals can afford to pass over the issues involved without discussion. Seabed manganese deposits called nodules should prove an important source of nickel, copper, and cobalt as well as manganese early in the next century.

During the past decade several international mineral combines explored for seabed nodules. The participating companies experimented with various processes for collecting the deposits from the ocean floor (2 to 6 km beneath the surface), and for extracting minerals once the ore is landed. (See table 6-1 for the principal consortia.) By 1982, the consortia had already spent an estimated $380 to $400 million on exploration and research. A number of technical problems remain to be solved before commercial production can begin, and there is no consensus on the prices necessary to make such production economically feasible.[19] Much will depend upon the future price of nickel, since an estimated two-thirds of the revenue will be derived from that mineral.

The scanty exploration for seabed nodules to date means there is little statistical basis for estimating their total volume.[20] The metal content of the nodules differs from area to area, but in the north and central Pacific, where initial ventures are likely to take place, the metal content (dry weight) is 18–24 percent manganese; 0.75–1.25 percent nickel; 0.50–1.15 percent copper; and 0.25–0.35 percent cobalt.[21] It has been estimated that an extraction unit should have a capacity of at least 3 million mt per year; this would yield about 35,000 mt of nickel; 30,000 mt of copper; 5,000 mt of cobalt; and 600,000 mt of manganese. A system of this size would require fixed assets in excess of $1 billion, and an overall investment, including working capital, on the order of $1.5 to $2.0 billion. One such operation could satisfy approximately 5 percent of 1979 world nickel demand; about 22 percent of demand for cobalt; about 30 percent for manganese; and about 0.3 percent for copper.[22] Two such operations could more than satisfy the United States' annual demand for manganese[23] and 150 percent of its annual cobalt demand.

Countries producing shoreside nickel, cobalt, copper, and manganese have been concerned lest world markets be swamped by this new source. However, given present technology, minerals produced from seabed nodules are likely to be of relatively high cost. Substantial increases in supplies from seabed nodules would quickly reduce world prices to the point at which production would become uneconomical. Of course, technological advances may reduce costs of seabed production, but advances in technology already reduce costs of shoreside minerals for the same grade of ore on a continuing basis. In one conceptual analysis of this problem, the general conclusion was

TABLE 6-1. International Seabed Mining Consortia

Participant	Parent Company	Country of Origin
Kennecott Consortium		
Kennecott Corporation	Standard Oil of Ohio	U.S.
RTZ Deepsea Enterprises	Rio Tinto-Zinc	U.K.
Consolidated Gold Fields	Rio Tinto-Zinc	U.K.
BP Petroleum Development	British Petroleum	U.K.
Noranda Exploration	Noranda Mines	Canada
Mitsubishi Group	Mitsubishi Corporation, Mitsubishi Metal, Mitsubishi Heavy Industries	Japan
Ocean Mining Associates		
Essex Minerals	U.S. Steel Corporation	U.S.
Union Seas	Union Miniere	Belgium
Sun Ocean Ventures	Sun Company	U.S.
Samim Ocean	Ente Nazionale Idrocarburi	Italy
Ocean Management Incorporated		
INCO	INCO	Canada
AMR	Metallgesellschaft, Preussag, Salzgitter	Germany
SEDCO	SEDCO	U.S.
Deep Ocean Mining Company	23 companies	Japan
Ocean Minerals Company (OMCO)		
Amoco Ocean Minerals	Standard Oil of Indiana	U.S.
Lockheed Systems Company	Lockheed Aircraft Corporation	U.S.
Ocean Minerals, Inc.	BKW Ocean Minerals	Netherlands
	Billiton B.V.	Netherlands

Source: Department of International Economic and Social Affairs, *Sea-Bed Mineral Resource Development* (New York: United Nations, 1982).

that "the impacts of seabed mining over the next 40 years on the welfare of land-based producers, consumers, and other groups range over an extremely broad spectrum, from negligible to overwhelming."[24] In other words, there is simply not enough information available to make a credible prediction at this time. However, it seems clear that even a few seabed mining operations could supply a large proportion of world demand for cobalt and manganese.

In addition to technical and financial considerations, uncertainty about eventual LOS mining regulations served to reduce undersea exploration and research after 1980. It was in response to this uncertainty that Congress passed the Deep Seabed Hard Minerals Resources Act of June 28, 1980. This act provides for government licensing of seabed minerals exploration and mining until such time as the United States becomes a party to an international treaty applicable to these activities. It provides for environmental regulations and a royalty of 3.75 percent of the market value of the minerals, to be placed in a fund that might be used for payments to an international authority set up by a treaty. Expressing the intent of Congress, the act sets forth conditions for the kind of international agreement the United States would sign. These include "assured and nondiscriminatory access, under reasonable terms and conditions, to the hard mineral resources of the deep seabed for U.S. citizens; security of tenure by recognizing the rights of U.S. citizens who have undertaken exploration or commercial recovery," and the absence of any restrictions imposing "significant new economic burdens" that might render exploration or mining unprofitable.[25] Any foreign nation that regulates the conduct of its citizens in a manner compatible with that provided in the act is to be designated a "reciprocating state." Special treaties might be negotiated with such countries to avoid conflicts over the exploration and commercial recovery of seabed nodules.[26]

Provisions of the LOS Convention Relating to Seabed Mining[27]

The LOS Conferences, which began in 1958, involved dozens of issues, with serious negotiations on seabed nodules beginning only during the Third UN LOS Conference, starting in December, 1973, and continuing through eleven sessions. At a meeting held on December 10, 1982, 117 nations signed the Draft Convention adopted the preceding April. The signers included most developing countries, the Soviet bloc, and all the developed countries except Belgium, West Germany, Italy, the United Kingdom, the United States, and Japan. When 60 states formally ratify the treaty embodying the April, 1982, convention, the program will go into operation.

The aspect of the convention of particular interest here is the declaration that seabed nodules are "the common heritage of mankind." To administer it, an International Seabed Authority (ISA) would be given broad powers over seabed mining, including approval of contracts with private or national governmental enterprises. Mining companies seeking a permit would have to explore two sites, providing full information on both sites to the ISA which would then have the option to develop one of them. The mining company might develop the other site under certain conditions and provided the contract to do so were approved by a majority vote of the ISA Technical Commission. The commission is to be elected by a three-fourths majority of an executive council seating representatives from thirty-six nations, each with one vote. (This council would be able to overrule the technical commission.)

The Draft Convention of April 30, 1982, foresees establishment of a supranational mining company called the Enterprise. Mining of sites explored by private or national companies is to be undertaken by the Enterprise without acquisition cost. The Enterprise would be guaranteed access to seabed mining technology owned by private companies applying for contracts, including that obtained by them under license from others. Similar access to privately owned technology is to be given any developing country planning to enter seabed mining. Such technology would have to be sold on reasonable terms to the Enterprise as a condition for granting contracts to private or governmental concerns. Financing for the Enterprise is to come from interest-free loans and assessments according to the scale used for regular contributions to the UN budget. (The U.S. share of UN contributions is 25 percent.)

The Draft Convention grants priority contracts to four international consortia that have engaged in seabed mining "pioneer" activities: the Kennecott Group, Ocean Mining Associates, Ocean Mining Incorporated, and Ocean Minerals Company (see table 6-1). Pioneer status entailing priority contracts is also provided to three state-sponsored programs in the USSR, India, and China. Contracts for at least one seabed mining site are guaranteed to each of these groups provided certain requirements are met, including the transfer of technology to the Enterprise and to other countries in accordance with the convention's terms.

There are both application and annual fees on the projected contracts, and contractors would pay either a production charge ranging from 5 to 12 percent of the value of the output, or a lower production charge (2 to 4 percent) plus a share of net profits. The net profits levy ranges from 35 to 50 percent during the first ten years of commercial production and from 40 to 70 percent during the second ten years.[28] Revenues from seabed mining, including those of the Enterprise and those paid to the ISA by contractors, are to be divided among some 150 nations expected to ratify the convention. (The convention also provides for shares to be paid to national liberation movements such as the Palestine Liberation Organization [PLO].)

The Draft Convention provides for production restrictions on seabed mining to protect the interests of land-based producers. Limitations may be affected by ceilings on output or by restricting the number of operations conducted by any one country. International buffer stocks would be used to stabilize prices of minerals from the seabed and to compensate countries suffering from falling prices. Thus the LOS convention envisages a comprehensive control of the industries producing the minerals contained in seabed nodules, whether from land-based or ocean sources. Such a program would clearly be unacceptable to the United States.

The convention stipulates that, after fifteen years of production, treaty provisions are to be reviewed to determine whether they have fulfilled policy objectives, among them protection of land-based producers, promotion of Enterprise operations, and equitable distribution of mining rights. Amendments to the treaty may be adopted after five years of negotiation; ratification by three-fourths of the signatory states would be required.

The U.S. Position on the LOS Draft Convention

Although neither the Ford nor the Carter administration was delighted by the prospects opened up by the convention intially drafted by Third World delegates, American policy succeeded in negotiating important compromises. This was especially the case while the American delegation to the LOS Conference was headed by Ambassador Elliot L. Richardson (early 1977 to October, 1980). The initial LDC position was that seabed mining would be conducted exclusively by the ISA, either directly by the Enterprise or by some kind of association between the latter and state or private corporations. The United States, of course, favored exploration and development by private enterprise. U.S. objections to the exclusivity foreseen for the ISA in Third World proposals led to tentative agreement on a "parallel" system of seabed mineral exploitation whereby state mining companies and private-sector corporations were placed on an equal footing with the Enterprise. Contracts would be granted by the ISA without regard to considerations other than financial standing and technological capability,[29] which proviso, of course, represented a retreat from the developing countries' postulate of international public ownership and development of the world's ocean resources. As a quid pro quo the U.S. delegation agreed to the provisions already outlined for providing the ISA and the Enterprise with the funds and technology they needed. Other issues were still being negotiated when, at the start of the tenth session of the Third LOS Conference, the Reagan administration undertook a major demarche.

In March, 1981, the president announced that the U.S. delegation would no longer participate fully in the conference until his administration had completed a review of the entire Draft Convention. Nearly a year later Reagan gave notice that the United States would seek basic changes in the

LOS draft with an eye to assuring development of seabed mineral resources adequate to national and world demand.

Most of the president's objections to the then-current Draft Convention were reflected in amendments put forth during the eleventh session of the conference (March and April, 1982). Together with six other industrial nations having a particular interest in seabed mining—Belgium, France, West Germany, Italy, Japan, and Great Britain—the United States proposed:

1. A guarantee that pioneer companies would obtain ISA authorization to mine;
2. An increase in size of the executive council majority needed for certain key decisions;
3. Provision that amendments to the seabed provisions require endorsement by all parties to the treaty (in effect an international *liberum veto*). Alternately, a two-thirds majority could pass an amendment, in which case the states opposing it would not be bound by the change (nonbinding amendments are not true amendments);
4. A requirement that the ISA mine sites be chosen by random selection rather than by its own perception of the better of two mine sites proposed by each contract applicant;
5. A guaranteed U.S. seat on the executive council; and
6. Less rigorous demands for technology transfer to the ISA.

A majority of the LOS Conference delegations objected to most of these amendments.[30] However, the final draft guarantees pioneering companies access to a specific mine site of up to 150,000 km^2. It also reserves a seat on the council for the "largest consumer" of seabed minerals—presumably the United States. Ambassador Richardson acknowledges that this assurance is somewhat ambiguous.[31]

After the Draft Convention was adopted in April, 1982, the United States negotiated an agreement with France, West Germany, and Great Britain to resolve any conflicting claims filed by seabed mining consortia, although all three countries reserved the right to ratify the Draft Convention. France has already done so; should West Germany and the United Kingdom eventually follow suit, this "minitreaty" might no longer be valid, since it is apparently inconsistent with the convention.[32]

The president of the LOS Conference threatened an International Court of Justice (ICJ) challenge to any nation violating the Draft Convention's provisions for seabed mining.[33] It is difficult to predict how the ICJ would rule on this question if a majority of the participant nations ratified the treaty. In any event, the United States might claim that the ICJ lacked jurisdiction. American corporations mining the seabed under permits issued by the U.S.

government might be provided with military security. Congress could also underwrite investment guarantees to protect U.S. companies from external interference. Without investment guarantees, U.S. mining companies probably would not risk the large sums required, nor could they obtain the necessary loans from commercial banks.

Controversy over the LOS

Ambassador Elliot Richardson, the most knowledgeable and articulate critic of the Reagan administration's rejection of the LOS Draft Convention, readily admits that the convention has undesirable features.[34] Nevertheless, he argues that the U.S. delegation should not have been instructed to vote against it; they should either have voted for it or at least abstained, as did several European countries.[35] Even if significant concessions were unobtainable, Richardson believes that (*a*) the convention's deficiencies are not so great as to make it unworkable as a treaty; and (*b*) the other provisions—those having to do with such matters as navigation, aircraft overflight, and fisheries—are extremely important to American interests: some of their advantages may be lost if the United States rejects the eventual treaty. Finally, Richardson holds that in the absence of U.S. treaty membership, domestic mining firms would have no legal right to exploit the seabed. He fears that in case of a dispute the ICJ would declare U.S. seabed-mining operations to be illegal.[36]

In the course of hearings before the House Committee on Foreign Affairs, Richardson engaged in a debate with James L. Malone, assistant secretary of state and the president's special respresentative to the LOS Conference. Malone believed that American mining firms would not engage in seabed mining under the treaty-to-be's terms, but might well do so under U.S. law and in the context of the minitreaties with other developed countries sanctioned by the Deep Seabed Hard Mineral Resources Act of 1980. Malone also argued that forced sale of technology to other countries would violate U.S. law and that the Senate would not ratify the convention in any event. Finally, Malone pointed out that the LOS nonseabed-mining provisions would be available to the United States even if it were not a signatory.[37]

The Reagan administration's decision against the LOS Draft Convention has been controversial in public forums as well as in the halls of Congress. The American Mining Congress and representatives of American consortia that have undertaken seabed exploration strongly opposed the proposed treaty.[38] The *Wall Street Journal* adamantly disapproved of the Draft Convention,[39] while the *New York Times* and the *Washington Post* were more favorably disposed to it. Sentiment for and against U.S. adherence was not divided neatly between liberals and conservatives; for example, an article in *Fortune* tended to favor the convention,[40] while an article in the *New Republic* (scarcely a spokesman for either the Reagan administration or the mining industry) quoted a former associate editor of the magazine.

> The phrase [common heritage of mankind] would be unobjectionable if it meant only that everyone has the equal right to go and stake his claim on a mineral site on the floor of the ocean. . . . Instead, the concept has been interpreted to mean that every nation has an equal right to the ownership of these resources, and to the financial proceeds from their exploitation. In other words, nations who spend nothing and risk nothing to extract the seabed resources have just as much say as those that spend millions of dollars in doing so. But until someone extracts the minerals, they are worthless. Whoever makes them valuable should reap the proceeds of what has been done. . . . The Reagan Administration took the correct approach to the negotiations by making clear that it would not sign a bad treaty. . . . The Administration should not endorse an ocean regime that offends this nation's philosophic principle at the same time it makes us poorer.[41]

Prospects for U.S. Ratification

The provisions of the Draft Convention work against efficient use of seabed minerals in several ways. First, the sites are not to be allocated to the entities best able to exploit them. Locations prospected and explored by mining consortia with the best available technology and the most experience are to be shared preferentially with the Enterprise or with developing countries, and Enterprise operations are to be heavily subsidized. Second, output limitations for the benefit of land-based producers would reduce productive efficiency: consumers throughout the world would be the losers. Third, high fees and profits taxes, together with production limitations and the method of allocating contracts, would substantially reduce (or eliminate) private investment in the industry.[42]

In considering whether the United States should ratify the LOS Treaty, these factors should be recognized:

1. The world is in no danger of exhausting shoreside resources of the minerals in seabed nodules for several decades.

2. Even under the most favorable economic and political conditions, actual production of seabed minerals is not likely for ten to fifteen years. Moreover, a great deal of additional research and experimentation is needed to decide whether seabed production can be as cheap as land-based output.

3. Uncertainty about the future of deep-sea mining under the LOS has at worst occasioned a hiatus for an industry that *may* assure adequate supplies of four minerals to last many generations. Alternatively, of course, tech-

nological advance could provide adequate substitutes for some or all of these minerals at prices lower than seabed mining requires.

4. As the major world consumer of minerals found in deep-sea nodules, the United States would gain from their availability regardless of who produced them.

The United States has two main options regarding the LOS treaty, both of which present certain disadvantages. It can ratify the treaty and seek to bring its implementation in line with U.S. policy. The disadvantage of this course of action is that it means agreeing to international policies and programs that contradict traditional American principles. Conceivably, some future administration will choose this option, but the chances of obtaining a two-thirds vote of the Senate for ratification are exceedingly slim. This course might speed up development of seabed mining, but it is doubtful that American firms associated with seabed mining projects in the past would participate under conditions established by the treaty.

Alternatively, the United States can reject the treaty and try to get around the pursuant disadvantages by such expedients as licensing American companies to undertake seabed mining outside the LOS agreement. The consensus of independent legal experts is that it would not be feasible for the United States to provide military security and investment guarantees for these firms.[43] Moreover, Congress might refuse to provide investment guarantees effective enough to circumvent the treaty. Perhaps a way around these problems of nonmembership would be to allow American companies to participate in consortia led by foreign concerns whose governments were treaty signatories. This alternative would provide a way of buying time until it is clear how seabed mining progresses under the LOS. Ambassador Richardson maintains that unless this country signs, it would lose advantages derived from the treaty's nonseabed provisions, but the Department of State holds that the United States would have these advantages in any case and that the nonseabed provisions accord with national policies. U.S. concerns would be unable to obtain ISA contracts to develop sites they had explored, but these firms could still profit through their membership with the consortia with ISA contracts. The United States could ratify the treaty at a later date, perhaps following amendments that would make it more palatable.

My tentative conclusion is that the last option is not only the most likely but also the most desirable under present circumstances. Given the treaty provisions that discourage private investment, it may make little difference whether the United States ratifies the agreement as presently constructed. Governments in Europe or Japan may decide to provide subsidies or guarantees to their mining firms that would make seabed mining under the LOS treaty attractive. United States consumers might well be the principal beneficiaries of such a development.

Is There a Basis for an International Agreement on Trade in Steel Products?

Chapter 5 reveals that trade in steel is encumbered by a maze of American and foreign import restrictions and "unfair trade practices" as defined by U.S. legislation and the General Agreement on Tariffs and Trade (GATT). Continuing disputes between the United States and the EEC, arising from imposition of penalty duties on U.S. products, have threatened to impair trade in a number of other commodities as well. United States penalty duties have also reduced steel imports from Japan and several developing countries.

In order to limit steel imports, the Reagan administration decided in 1984 to negotiate VRAs with all steel-exporting countries. Under this arrangement, each exporting country would limit its sales to a certain percentage of the U.S. market. The adoption of similar programs by other industrial countries would create a system of managed international trade in steel, which would virtually destroy international steel competition, raise prices to consumers, and delay modernization and adjustment of the American and European steel industries. Import prices as well as domestic prices would rise, since foreign producers would have no incentive to compete for a larger share of the American market.

Are there other types of international agreements that would better serve U.S. and world consumers and promote productivity in the steel industry? An ideal solution would be an arrangement whereby major industrial countries eliminated both import restrictions and unfair trade practices in steel and submitted all disputes to the GATT for resolution. Such an agreement (covering other commodities as well) already exists among the contracting parties to the GATT—all the major non-Communist exporters and importers of steel products. But the GATT has not worked well in the area of nontariff barriers for several reasons. First, the nature of unfair trade practices in steel products is subject to dispute. For example, some countries argue that they should be able to meet competition from third countries in U.S. markets even if this means selling below prices in their home markets. There is also disagreement over the definition of export subsidy. Second, the parties differ as to the types of import restriction legitimated by serious injury from imports and as to how exporting countries should be compensated for these restrictions. Third, the GATT secretariat has no power to actually settle disputes; it can merely provide a forum for the reconciliation of differences.

Resolution of trade problems in steel would probably require a special accord on a number of complex issues and an international agency with adequate power to implement it. Such an accord should include precise definitions of trade practices in steel; specific limitations on temporary import restrictions arising from injury to domestic producers; and the acceptance of agency decisions in disputes occasioned by implementation of the agreement.

The logical agency to sponsor an international conference leading to such a pact, including provision for its administration, is the GATT. An agreement on steel along these lines would probably require approval by Congress.

A full discussion of what a sectoral accord on steel would involve is beyond the terms of reference of this study.[44] Such an arrangement is, I believe, both possible and desirable—if the United States and other industrial countries want to negotiate it. The agreement could also be applied to exports of developing countries.[45]

International Cooperation for Dealing with Import Disruption

One criticism of the present U.S. strategic stockpile program is that it pays little or no attention to the requirements of America's allies and other friendly countries in the event that the supply of one or more nonfuel minerals is disrupted. In case of war it is likely that the United States would assist its NATO allies and Japan in meeting their mineral requirements. The International Energy Agency (IEA) has established a program for the sharing of petroleum, including use of strategic oil reserves if supply is greatly disrupted. Similar arrangements might be established for critical and strategic nonfuel minerals.[46]

Supply Disruptions in Peacetime

Since all countries would share in the benefits of stockpile releases under the program outlined in chapter 7, a scheme for international cooperation among developed countries to deal with the interruption of nonfuel mineral supply in peacetime would be highly desirable. Such a program should be limited to developed countries rather than including LDC producers of these commodities because an economic stockpile's sole raison d'être is to protect consumers from import disruption. To include Third World producer countries in the program might introduce broader objectives, such as the maintenance of price floors.

International cooperation of this sort among major non-Communist industrial countries should be based on three principles. First, governments should agree not to impose export restrictions or otherwise interfere with international nonfuel mineral markets. Second, programs to maintain private or government inventories of designated minerals should be established. Third, plans should be coordinated for stockpile releases in the event of a global supply disruption. These releases would take place only in response to previously designated types of supply disruption in the source countries that resulted in a specific percentage reduction of world supplies, or a specific percentage rise in world prices. The nonfuel minerals selected for inclusion should be limited to those for which there is a significant probability that supply over the next two or three decades might be severely curtailed.

A multinational cooperation scheme of this type would have certain advantages over national programs. A large part of U.S. imports of nonfuel minerals are supplied in processed form by industrial countries that import raw materials from developing countries—where supply reductions are most likely to occur. This pattern of specialization and trade should be protected against sudden unavailability of raw materials. The maintenance of global inventories as contrasted with national inventories would involve important economies by avoiding duplication. Finally, no single country would gain an advantage from global supply disruption simply because it was better endowed than another in a particular mineral.

Admittedly, such a cooperative program would be difficult to establish and equitably administer. The record of multinational economic cooperation among industrial countries has not been favorable in recent years. Nevertheless, the threat of global supply disruption provides a focus for cooperation that just might succeed, and it is important enough to be placed on the agenda for the annual Economic Summit Conference. An agreement among non-Communist industrial countries to maintain free trade in nonfuel minerals would make it unnecessary for EEC countries and Japan to accumulate stockpiles of commodities such as copper and iron ore for which they depend heavily on imports. Important producers such as Australia, the United States, and Canada would agree to maintain unrestricted trade in these commodities, except possibly for exports to Soviet bloc countries. The EEC and Japan could therefore better afford to cooperate in a multinational program to stockpile minerals such as cobalt and chromium, the global supplies of which are highly vulnerable to disruption.

Conclusions

Successful international agreements on economic issues require compatible national objectives and similar economic systems among parties to the agreements. These conditions exist to a significant degree among developed countries, but not between them and developing or Communist nations. Therefore, international approaches to such issues as commodity price stabilization, mobilization of capital for development of Third World minerals, and mining of seabed minerals are less likely to be effective than multinational cooperation among industrial countries to deal with unfair trade practices or import disruption. The United States and certain other major industrial nations do not have the same objectives for price stabilization of nonfuel minerals as major LDC producers of these minerals. The latter countries have been seeking an international mechanism, including export and import controls, to support world prices of nonfuel minerals. The United States has never been enthusiastic about commodity price agreements but in the past has been willing to cooperate in buffer stock arrangements that seek to avoid sharp movements above and below long-term equilibrium price trends. Yet agreement on targets and the methods by which targets are realized has been exceedingly difficult to achieve. Moreover, the conclusion of satisfactory accords may be thwarted by the need to accumulate unacceptably large amounts of both commodities and financial resources.

The approach to seabed mining that would combine private or state mining operations under international agency control with parallel mining activities by the agency itself has foundered because of differing national objectives, economic structures, and ideologies. The United States and certain other industrial countries have not accepted the principle that minerals in the oceans constitute the "common heritage" of all national states. Seabed minerals lying several miles under the ocean have no value unless billions of dollars are spent to find, recover, and process them. Industrial nations with the capital and skills necessary to carry out these activities are reluctant to subject themselves to control by an international agency dominated by Third World states. Therefore, I foresee little progress in making seabed minerals available to world consumers under the LOS Treaty as presently drafted.

The use of World Bank and other international assistance agency resources to finance nonfuel mineral development in LDCs also brings the national objectives and economic structures of the members of these organizations into conflict. Third World countries have viewed development agency financing as a means to support state mining enterprises and, in some cases, as a means to develop their mineral resources without assistance by multinational mining corporations. The U.S. mining community has opposed direct and indirect international assistance agency aid to enterprises as a form of subsidy to its competitors. The U.S. government has on several occasions voted against loans by the World Bank Group and the IADB to state mining enterprises, favoring modest loans and equity investments by public international agencies as a catalyst for private investment in Third World mining. In view of the international development agencies' limited resources and of the desirability of a larger flow of capital and technology to mineral industries in Third World countries, I believe this is the correct approach.

The issue of public international development assistance to Third World industries that compete with products of the United States and other industrial countries in world markets encompasses much more than nonfuel minerals alone. International development assistance is provided directly and indirectly to a wide range of Third World industries that compete in international markets. Economic development requires that exports of both raw materials and industrial goods be expanded, so the World Bank Group and similar development assistance agencies should not be forbidden to promote exports from the developing countries. In the case of nonfuel minerals, however, international private capital is available for mining and processing provided investment conditions are attractive: international public capital should be reserved for development areas where private capital is unobtainable.

Differences in broad national policies do not complicate the problem of remedying the chaotic conditions in steel trade among industrial countries. All industrial countries profess to support the fair trade provisions of the GATT. The basic problems here are the absence of statemanship in the area of international trade and the inability of governments to subordinate the special interests of broad sectors of the economy to national welfare. Current U.S. policies are leading in the direction of a managed international steel trade, a development that has costly implications for world consumers of steel and will delay modernization of the steel industry. A more acceptable approach would be the negotiation of a sectoral agreement in steel products to be administered by the GATT, which should be strengthened to deal adequately with disputes arising under such an accord.

International cooperation among developed countries to forestall potential nonfuel mineral import disturbances is not stymied by opposing national objectives and would appear to be quite feasible. The obstacles are largely inertia and perhaps different views on the nature of vulnerability. An international cooperation scheme exists for petroleum, and it is highly desirable that industrial countries establish coordinated programs to handle possible nonfuel mineral supply disruptions.

7

Measures to Reduce Vulnerability to Import Disruption

For the past several decades the United States has maintained a stockpile of imported materials to assure their availability in case of wartime emergency. The likelihood of emergencies such as those the stockpile aims to meet and the operational features of the program itself have both been met with skepticism. A major criticism is that the program is oriented to cope with shortfalls arising from a conventional war of several years duration rather than to more likely interruptions due to limited wars or civil disturbances abroad. This chapter reviews briefly the history of the U.S. strategic and critical minerals stockpile program. Following an examination of likely sources of foreign supply disruption, this chapter outlines a scheme to reduce American vulnerability in circumstances other than those occasioned by a major war involving the United States. Finally comes an analysis of alternative measures—subsidized private inventories and production, for example.

The U.S. Strategic Stockpile Program

America's effort to create inventories of commodities needed in the event of war began with the Strategic and Critical Materials Stock Piling Act of 1939 followed by the more comprehensive legislation of the Strategic and Critical Materials Stock Piling Act of 1946 and subsequent revisions. The main purpose of the program, according to a presidential report of April 5, 1982, is to provide a stockpile adequate for military, industrial, and essential civilian needs during the first three years of a major conventional war.[1] During its long history, the U.S. strategic stockpile program has changed its goals from time to time according to the government's varying wartime scenarios. A five-year war, patterned in considerable degree on the experience of World War II, was the assumption that determined initial goals under the 1946 Stock Piling Act. In 1958 the planning basis was reduced from five to three years of warfare, and in the early 1970s new strategic assumptions resulted in a sharp lowering of inventory requirements, thus automatically creating large surpluses of a number of materials. Despite the sale of over $5 billion worth of excess materials in the 1960s and early 1970s, at the end of 1972 some $2.4 billion of the total $6.4 billion in government inventories exceeded objectives.[2]

President Nixon believed that the future war assumptions on which stockpile goals were based were unrealistic: in April, 1973, he asked Congress for authority to reduce the inventory by 90 percent. Nixon's program involved preparation for only a one-year war, so objectives were reduced in 1973 by more than $4 billion, leaving the goal at only $700 million in 1973 prices. However, this reduction was never achieved because it takes considerable time to shrink inventories without seriously affecting commodity prices.

The Ford and Carter administrations substantially increased stockpile goals. These now assumed a three-year emergency with priority to a one-year NATO war in Europe, fought with either conventional or nuclear weapons but involving no direct nuclear exchange between the United States and the USSR. This remained the basic strategic assumption for the Reagan administration.[3] Under Reagan, a three-year mobilization plan was formulated by the National Security Council to attain defense objectives while maintaining essential civilian industries. This plan provides the basis for determining stockpile goals.[4] It involves rationing of certain commodities, price controls, investment and housing construction targets and controls, import and export regulation, and stockpile releases to civilian and defense industries consistent with mobilization plan priorities. There is no indication that prices might be used to limit effective demand either for critical materials or for civilian consumption.

Once requirements for the three-year war scenario are determined, supplies of nonfuel minerals from foreign sources are estimated in keeping with assumptions about the accessibility of those sources in time of war. Canada and Mexico are considered assured suppliers, while each of the remaining sources is rated according to political orientation toward the United States and ability to sustain exports of stockpile materials in wartime. Transportation loss estimates are also weighed in determining the practical availability of foreign supplies. Imbalances of requirements over supplies are then estimated for each year of the war emergency and a preliminary stockpile goal is calculated for each material. Requirements are divided into three sectors, namely, defense, essential civilian, and general civilian, for each year of war. Defense and essential civilian needs have the higher priorities in determining stockpile objectives. The Federal Emergency Management Agency (FEMA) sets the inventory goal for each material according to wartime policy targets established by an interagency committee that includes the Departments of Defense and Commerce.

The present National Defense Stockpile Inventory of Strategic and Critical Materials (NDS) consist of sixty-one materials or material groups. About fifty of that total are nonfuel minerals, which also constitute the vast bulk of the inventory's value—they were worth $11.1 billion as of September, 1983. (See app. 7-1 for a list of all the materials involved.) Because goals are changed from time to time, some materials in the inventory, about $4.2 billion worth, are in excess of present objectives. Some, such as silver and columbium metal, are no longer on the list at all. Existing stocks of others, such as bauxite, cobalt, tantalum, and zinc, are below current goals; acquiring enough of them to meet the goals would cost $9.8 billion in 1983 prices. However, current legislation (1979 Stockpiling Act as amended) rigidly circumscribes both disposal of excess inventory and purchase of stocks: it will probably be many years before superfluous reserves can be liquidated and additional inventories purchased.[5] In the meantime, goals are likely to change, so that under present legislation it is difficult to see how actual inventories could ever be balanced with goals. As an example of the changing goals of the national defense stockpile and the changing perception of stockpile requirements, in July, 1985, President Reagan proposed that stockpile goals be reduced from $16.3 billion to $6.7 billion, and that no additional *net* purchases of materials for the stockpile be made. The new stockpile goals would eliminate $9.7 billion in unmet goals under the program as of June, 1985, and leave $3.2 billion as surplus materials above the new goals, of which $2.5 billion would be sold over the next five years. Receipts from sales would go to fill unmet materials goals, with the remainder used to reduce the federal budget deficit. Despite the proposed 59 percent reduction in stockpile goals, the White House announcement stated that the revised goals "will be sufficient to meet military, industrial, and essential civilian needs for a three-year conventional global military conflict, as mandated by Congress in 1979." According to the president's proposal, inventories of strategic materials valued at only $0.7 billion would be required during a protracted military conflict for supplementing materials not available in sufficient quantities from domestic or reliable foreign sources. The stockpile would, however, contain a "supplementary reserve" of materials currently valued at $6 billion.[6] This proposal, which requires Congressional approval, has met considerable opposition from the mining community.

Although releases of materials from the NDS for use in civilian industries are not permitted except in times of national emergency, the president can authorize them for the exclusive use of defense industries. If the administration authorized releases for defense industries threatened by severe import disruptions in peacetime, civilian industries would have increased access to supplies for those materials because all or part of defense requirements would be drawn from the NDS.

Criticisms of the National Strategic Stockpile Program

One of the most compelling strictures directed at the strategic stockpile program is that it is predicated on an unrealistic and rigidly defined war scenario, including a war mobilization plan that establishes arbitrary levels of civilian production and consumption for various economic sectors. Although it is conceivable that World War III might be fought almost entirely in Europe, Asia, or Africa with little damage to the American economy—as was the case during World War II—many regard this as highly unlikely in the nuclear age. Much more likely is either a series of limited wars in which the United States may or may not be directly involved, or a short intercontinental nuclear war during which stocks of finished goods at appropriate locations would be more useful than raw materials for a highly complex industrial economy. In other words, the present stockpile program is not designed to deal with the most likely contingencies.

The current stockpile program is often faulted as both expensive and inefficient. Because inventories can be adjusted to changes in military and political outlook only over long periods of time, there are large excess stocks of certain commodities and deficits in others. In addition, a substantial portion of the existing stockpiled materials has been allowed to deteriorate. Only recently has there been a serious effort to actually determine the quality of materials accumulated over the past thirty years.[7] Physical administration of the stockpile has been so poor that it is questionable how much of it could actually be used in an emergency.

A related complaint is that the materials are frequently not in forms utilized by industry, especially since the required forms are constantly changing. For example, different grades of bauxite are used to make alumina or in such employments as abrasives; there are chemical, metallurgical, and refractory grades of chromite ore; and there are chemical-, metallurgical-, and battery-grade manganese ores. Since America does not have enough capacity to process several nonfuel mineral ores into the forms required by industry, certain minerals ought to be held in the appropriate processed state. For instance, the United States lacks sufficient capacity to transform enough bauxite into the alumina for its aluminum metal industry needs. Therefore, alumina or aluminum ought to be held in reserve as well as bauxite. Much the same holds true for U.S. production of ferroalloys from chromium and manganese. For some minerals, American processing capacity has been declining and a larger proportion of imports has thus been in semifinished shape. A strong argument can be made, therefore, that government programs should be directed at enhancing the stocks of critical minerals in the forms they are actually used and by the concerns that actually use them, rather than through the cumbersome NDS mechanism. Mining representatives are foremost among those who point out that the erratic nature of stockpile operations in the past has had an adverse effect on their industry and on the creation of additional domestic productive capacity. Large amounts of certain materials have been accumulated only to be later sold, often within a relatively short period of time.[8] Although the 1979 Stockpiling Act reduced the power of the president to dispose of inventory without consent of Congress, legislative supervision also involves a certain politicization. For example, in 1981 congressional pressure caused sales of silver from the stockpile to be indefinitely suspended, although large existing stocks were no longer essential to national security and even though the high price for silver in 1981 provided a good opportunity to sell it.

A final criticism of the program is that goals are set without evaluating government expenditures according to social benefit-cost criteria. Objectives for civilian requirements are now established strictly on the basis of projected levels of consumption (adjusted for wartime conditions). This takes no account of the relationship between the social costs of maintaining a certain inventory of a material and the corresponding social benefits, or of the trade-off against social benefits that might have been derived from alternative uses of public funds.

The present national strategic stockpile program can be justified only if there is sufficient probability of a long war that would disrupt most overseas supplies of materials, and the domestic production system would not be subject to severe damage. This probability must be high enough to warrant the use of public funds for the stockpile over other possible defense uses, such as accumulating essential military and civilian finished products. This is a matter of judgment for which we have little empirical basis, but the program's history offers no evidence that a serious weighing of probability-adjusted social benefits and costs has taken place in recent years. Even in terms of the program's own objectives, its management as established by legislation is severely deficient. This deficiency was recognized by the National Strategic Materials and Minerals Program Advisory Committee when it recommended that the NDS program be administered by a government corporation. This suggestion was based on the committee's finding "that efficient, sound, businesslike management of the stockpile is difficult under the present system in which decision-making authority is so widely distributed among disparate agencies with correspondingly disparate 'missions.' "[9]

Following the 1973–74 Arab oil embargo, some members of government advocated a broadening of the strategic stockpile program to include peacetime releases for general industrial needs when supply was disturbed.[10] Congressional interest in the availability of a stockpile during peacetime is reflected in the numerous hearings on the subject and by a study on economic stockpiling published by the Office of Technology Assessment at the request of the House Science and Technology Committee.[11]

Combining the functions of the present NDS with those of a stockpile available for nondefense purposes in peacetime would cause problems. First, the size and composition of the national stockpile would need to be altered to accommodate both functions. The nonfuel mineral stocks required for an all-out war contingency would have to be supplemented by an additional amount necessary to avoid or mitigate the economic impact of an import disruption of each commodity in peacetime. Second, under current regulations NDS releases are to be made under a system of government allocation and price controls, which is not appropriate for a peacetime economy. Third, for an economic stockpile to be cost effective, the criteria for acquisitions and releases should be quite flexible and exempt from the severe legislative constraints that currently circumscribe NDS operations.

Vulnerability in Peacetime

The nonfuel minerals subject to import disruption in peacetime are not those produced by the United States and other major industrial powers—metals like aluminum, steel, and refined copper (see chap. 5). These industrial products experience price increases from time to time as a consequence of world demand and supply conditions (including reduced output in some countries due to labor disputes), but this is not abnormal in the world economy.[12] The materials at risk are mineral ores such as bauxite and manganese, and minor metals such as cobalt and chromium, whose production is heavily concentrated in developing countries, in the Soviet bloc, or in South Africa.

Limited Wars and Revolutions

Highly probable sources of supply disruption are limited wars and revolutions in countries or regions with great actual or potential political instability. Major attention has focused on central and southern Africa because they are significant producers of several important nonfuel minerals. But these are not the only areas where disruption from war or revolution is possible. Jamaica and Suriname, which are important sources of bauxite and alumina, have recent histories of political instability, while Thailand, a major source of tantalum, is threatened by Vietnamese troops over its northern border.

Cartels

Cartels strong enough to double or even triple the world price of a mineral may cause another type of supply difficulty. In order for a cartel to obtain monopoly profits by restricting output for several years, it must control a fairly large proportion of world output, at least 50 percent; and both world demand and supplies from other sources must be relatively inelastic. Appendix 7-2 gives an analysis of a cartel's ability to increase profits by restricting output.

The elasticity of both world demand and noncartel supply is likely to be quite low for most minerals in the short run, but each tends to rise over time. The noncartel suppliers can develop increased capacity or expand output with existing capacity. Thus, for most minerals, the period during which a cartel can maintain very high prices by restricting output is likely to be short, perhaps no more than three years. Moreover, expansion of noncartel sources may permanently shrink the world market share held by cartel members, and the use of substitutes could reduce the size of the market itself. These possibilities serve as a deterrent to cartel action, when short-term monopoly profits are to be made.

Cartels generally require government support; without it they are likely to run afoul of anticartel laws in the United States and Western Europe. Discipline is hard to enforce since any member could take advantage of the other members by increasing its output or sales beyond its quota. The OPEC cartel was successful during the 1970s largely because one member, Saudi Arabia, was willing to cut back its production while most of its partners produced as much as they desired. When world demand for petroleum declined in the 1980s and Saudi Arabia was no longer willing to further cut back its production, the world price of oil declined. By the mid-1980s OPEC was having great difficulty upholding the cartel price because some OPEC members were exceeding their output quota or were selling at discounts.

Among nonfuel minerals producers none has the financial resources of Saudi Arabia. The political and economic preconditions do not exist for a nonfuel mineral cartel whose members could be trusted to adhere to agreed-upon output or export quotas.

Politically Oriented Embargoes

Most nonfuel minerals have diverse sources of supply: countries with a variety of political orientations and political and economic relationships with the Western industrial powers. It is difficult to imagine any combination that might initiate a politically inspired stoppage of shipments. The Arab oil embargo touched off by the 1973–74 Arab-Israeli war resulted from temporary political cohesion among a group of major suppliers of a resource vital to Western economies, but no other commodity offers a parallel. Moreover, the Arab action clearly showed that a selective embargo (e.g., against the United States and the Netherlands) could not be enforced. There are a few nonfuel minerals, such as cobalt or chromium, whose production is concentrated in two or three countries, but an export embargo on these commodities is neither politically nor economically feasible.

A Resource War

In recent years some writers have raised the specter of a resource war between the USSR and the Western powers. The resource war scenario has dramatic appeal, but little if any analytical content: it is based on unrealistic assumptions.[13] It hypothesizes a rapidly growing world shortage of resources and consequent competition for supplies reminiscent of the nineteenth-century colonial rivalry of European powers. Those who envisage a resource war say that sources of minerals (including petroleum) vital to the Western world might come under the control of the Soviet Union, which would use this control for political blackmail, or to reduce the defense capability of the United States and its allies. It is apparently assumed that such hegemony could be achieved by fomenting revolutions that established Communist governments subservient to the USSR. (Certainly, any attempt to achieve this objective by external military invasion would spark a world war involving the United States and other NATO countries.)

The idea that the USSR could succeed without military conquest in promoting the establishment of subservient Communist regimes throughout central and southern Africa and the Middle East is inconsistent with recent historical experience. Even Marxist regimes are highly nationalistic and try to maintain their independence. Granted that Russia has been seeking (with rather poor success lately) political influence in Africa and the Middle East, it is difficult to see why it would want to divert the minerals of these areas from world markets to itself. The Soviet Union is self-sufficient in most nonfuel minerals and exports several to the West, including chromium and the platinum-group metals. Despite close ties to the Soviet Union, countries like Angola are eager to sell their petroleum, and diamonds, and other resources in world markets. The USSR and its European satellites must also sell whatever minerals they can spare to get convertible currencies to pay for their imports and to service their large debts.

Criteria for Assessing Vulnerability in Peacetime

Some risk of foreign supply disruption is characteristic of an interdependent world economy, and every nation is vulnerable to disruptions in world trade from a variety of causes. What concerns us here is the *degree* to which supply of a particular mineral is subject to interruption. Such an assessment is required before deciding whether special governmental action is warranted to counter a supply disruption. There is no broad consensus on criteria, but the following considerations may suggest some guidelines.

1. A disruption should measurably affect GNP and the balance of payments or damage the economic welfare of a significant percentage of consumers, either by reducing the quality of finished goods or by limiting consumer choice. A material used for a small volume of nonessential products of special interest to less than, say, 10 percent of the population should not qualify.

2. Disturbance in trade for a mineral should increase its price several-fold over the predisruption level. A rise of up to 50 percent frequently occurs as a consequence of normal market developments. The extent of the price rise after a disruption depends upon both the percentage of normal world supply that is restricted and the elasticity of world demand and supply. The lower these elasticities the higher will be the rise in world price with a given reduction in supply.[14]

3. The imported mineral should have no readily available substitutes for use in production of finished goods.

4. Exposure to import disruption is greatest when world supply is concentrated in geographical areas themselves subject to social and political disturbance. Such is the case for Zaire and Zambia, where global cobalt sources are highly concentrated. On the other hand, there are a number of bauxite- and alumina-producing countries, including politically stable nations like Australia. As a consequence, political instability in one source (e.g., Ghana) does not much increase the risk associated with importing these materials used to make aluminum metal.

Experts differ as to the list of nonfuel minerals so vulnerable to import disruption as to warrant public action. Some would include no more than a half-dozen—usually bauxite-alumina, chromium, cobalt, manganese, and the platinum-group metals. (App. 7-3 provides a detailed analysis of the five minerals or minerals groups most liable to disturbance or cutoff.) United States net import reliance on imports for this group ranges from 88 to 98 percent, and except for bauxite-alumina, production outside the Communist world is heavily concentrated in one or two countries (see table 7-1).[15] Others would add columbium, industrial diamonds, tantalum, and tin; still others would even include minerals that the United States produces in sizable quantities, such as lead, silver, tungsten, and zinc. These disruptions reflect different perceptions of the probability of import disruption, and different methods of estimating costs. However, few if any serious students of the problem would include all of the materials now in the national strategic stockpile.

To judge by the above criteria, most of the twenty-six important minerals listed in table 7-1 pose no serious threat to the American economy via import disruption. One or more of the following holds true for most of them: (*a*) import dependence is minor and could easily be lowered by expanding production or by conservation; (*b*) there are readily available substitutes for most uses; (*c*) supply sources are well dispersed, mainly in politically stable developed countries.

For five of these commodities—aluminum, copper, iron and steel, lead, and molybdenum—the United States is either a net exporter or usually has a net import balance of less than 15 percent of consumption. The import bal-

TABLE 7-1. U.S. Net Import Reliance on Selected Nonfuel Minerals as a Percentage of Apparent Consumption, 1980 and 1981

Metal or Mineral	1980	1981	Major Foreign Sources, 1977–80 (average)
Aluminum	E	E	Canada (62%), Ghana (11%)
Antimony	48	51	Metal: China (39%), Bolivia (29%), Mexico (12%) Ore: South Africa (48%), France (14%), China (13%), U.K. (12%)
Bauxite and alumina	94	94	Bauxite: Jamaica (45%), Guinea (26%), Suriname (13%) Alumina: Australia (74%), Jamaica (16%), Suriname (8%)
Chromium	91	90	Chromite: South Africa (44%), USSR (15%), Philippines (16%), Finland (9%) Ferrochromium: South Africa (71%), Yugoslavia (11%), Zimbabwe (7%), Sweden (4%)
Cobalt	93	91	Zaire (42%), Belgium-Luxembourg (15%), Zambia (14%), Finland (7%)
Columbium	100	100	Brazil (74%), Canada (6%), Thailand (6%)
Copper	14	5	Chile (29%), Canada (23%), Peru (14%), Zambia (12%)
Diamonds (industrial)	100	100	Ireland (38%), South Africa (31%), Belgium-Luxembourg (6%), U.K. (4%)
Gold	18	7	Canada (40%), USSR (25%), South Africa (17%)
Ilmenite	35	43	Australia (64%), Canada (29%), South Africa (4%)
Iron ore	25	28	Canada (65%), Venezuela (16%), Brazil (9%), Liberia (6%)
Iron and steel	15	14	Japan (37%), Europe (37%), Canada (12%)
Lead	E	10	Canada (28%), Peru (20%), Honduras (25%), Australia (13%)
Manganese	98	98	Ore: Gabon (40%), Brazil (19%), Australia (15%), South Africa (14%) Ferromanganese: South Africa (39%), France (25%)
Mercury	26	39	Spain (33%), Algeria (17%), Japan (15%), Italy (10%)
Molybdenum	E	E	Canada (92%), Chile (3%)
Nickel	73	72	Canada (46%), Norway (10%), Botswana (8%), Australia (6%)
Platinum group	88	85	South Africa (55%), USSR (18%), U.K. (11%)
Rutile	W	W	Australia (81%), India (5%), Sierra Leone (5%)
Silver	7	50	Canada (40%), Mexico (24%), Peru (22%), U.K. (5%)
Tantalum	90	91	Thailand (36%), Canada (11%), Malaysia (11%), Brazil (5%)
Tin	79	90	Malaysia (44%), Thailand (18%), Bolivia (17%), Indonesia (11%)
Titanium	W	W	Japan (74%), USSR (14%), China (9%), U.K. (3%)
Tungsten	53	52	Canada (30%), Bolivia (24%), China (11%), Thailand (10%)
Vanadium	17	42	South Africa (58%), Chile (16%), Canada (7%)
Zinc	60	67	Ore: Canada (50%), Peru (16%), Honduras (10%) Metal: Canada (49%), Spain (8%), Mexico (7%), Australia (6%)

Source: Bureau of Mines, *Mineral Commodity Summaries 1982* (Washington, D.C.: U.S. Department of Interior, 1982), pp. 4–5.

Notes: Net import reliance = imports minus exports plus adjustments for government and industry stock changes. Apparent consumption = U.S. primary plus secondary production minus net import reliance. E = net exports; W = withheld.

ances for gold and silver fluctuate widely in response to nonindustrial demand and supply. Iron ore is extremely important to the economy, but the United States has reserves and shut-down capacity that could be brought on stream fairly rapidly. Nearly two-thirds of American iron ore imports come from Canada.

Of the remaining eighteen nonfuel minerals listed in table 7-1 for which the United States has a net import balance of over 25 percent, eight can readily be replaced by substitutes in most uses.[16] Imports of four of these minerals come largely from Canada, while import sources for several others are widely diversified among developed and developing countries. Canada, Norway, and Australia supply over 60 percent of U.S. nickel imports. There are no satisfactory substitutes for titanium (sponge) metal for aircraft and space uses. The United States imports more than half its consumption of titanium ores (ilmenite and rutile), mainly from Australia and Canada. It is more than 80 percent self-sufficient in both titanium metal and titanium dioxide (mainly used in paint and paper). Net mercury imports as a percentage of consumption have varied widely in recent years—from 59 percent in 1979 to 26 percent in 1980—and there are no satisfactory substitutes for it in electrical apparatus and industrial control instruments. However, over half of U.S. imports come from developed countries, principally Spain, Japan, and Italy.

For minerals that are used heavily in defense industries, a severe cutback of imports from an important source is likely to result in the release of materials from the strategic stockpile.[17] This applies particularly to imported alloy metals such as columbium, tantalum, tungsten, and vanadium. Such releases would also reduce disruption for the nondefense industries that use the minerals.

The Economic Stockpile

It is clear that the current strategic stockpile is inadequate as a hedge against damage to nondefense industries dependent on a few critical minerals susceptible to import cutoff. It seems prudent therefore to create a specifically civilian program, designed for peacetime use, to meet this need.[18] The size of an economic stockpile inventory for each material should be determined by comparing the social cost of stocking a given amount with the social benefit from releasing that amount in case of a foreign supply stoppage. The social benefit would equal the reduction in social loss that would occur without an inventory release. Since no one knows when, if ever, an import disruption will occur, potential social benefit must be adjusted for the probability of its occurrence. If the social benefit from release of a million tons of a mineral is estimated at $100 million, and the probability of an import disruption over the stockpile program's duration is 30 percent, the probability-adjusted social benefit is $30 million. Therefore, the social cost of stocking a million tons of a mineral over the planning period should not exceed $30 million. It would be economical to accumulate still more of the material so long as the social benefit expected from its release in the proper circumstances exceeded the additional cost of the larger stockpile.

But why not stock enough of each mineral to offset any conceivable loss from an import disruption? The very high social costs of such a course would probably outweigh its social benefits when the latter were discounted by the merely 30 percent probability of realization over, say, a ten-year period. Expenditures to stockpile critical materials might be regarded as insurance payments for protection against the aggregate of possible social losses. Just as with conventional insurance, premiums should be adjusted upward for contingencies with the greatest probability. In this manner most losses could be covered without spending more than a fraction of the aggregate loss were all potential supply cutoffs to occur.[19] This optimum result would require that the probabilities for a large number of contingencies be correctly determined.

Stockpile releases should be sold at the world market price and not rationed. Rationing is incompatible with a peacetime market economy; it would inevitably require a complex system of price controls and allocations that would impair adjustment to import disturbances. It is also important that releases be made only in the event of substantial foreign supply disruptions that would raise world prices by 100 percent or more: an economic stockpile should not be used to counter price fluctuations arising from normal shifts in demand and supply. Finally, an economic stockpile should involve no government restrictions on either exports or imports designed to influence world market prices. Restrictions on exports would deny the benefit of the lower prices resulting from the stockpile releases to our allies and other friendly countries, while restrictions on imports would reduce the incentive for non-disrupted foreign sources to expand output.

The Social Benefits from Stockpile Releases

If stockpile releases were sold on the market, their principal effect would be to moderate the rise in world price caused by an interruption of foreign supply. By lowering the world price from the level it would otherwise have attained, a release could avoid or at least reduce social losses—and thereby produce social benefits.

The most important amelioration of loss made possible by stockpile releases would be the reduction in additional revenues going to foreign suppliers due to the increased world price occasioned by import disruption. (It should be noted that disruptions can scarcely be more than partial because it is highly unlikely that all foreign supply sources would be simultaneously closed off.) If the imported material were not also produced domestically, consumers would lose an amount equal to the difference between the world price *before* and *after* import disturbance, multiplied by the import volume afterward. Consumers would also lose because they could buy less of the imported material at the higher price.[20] Stockpile releases would produce their greatest social benefits by reducing or eliminating these two losses.

If there is some domestic production of an imported mineral (or if there is a substitute for it), the rise in world price caused by an import disruption may induce more domestic output at a higher cost. The additional real cost of such production (compared with the previous cost of importing the same amount of the mineral) also constitutes a social loss. Thus a stockpile release could also generate a social benefit by making some or all higher-priced domestic production unnecessary.

A further social loss results if the higher price occasioned by import disruption causes a temporary loss of output while retooling goes on. For example, a multiple price rise for the cobalt used to manufacture magnets might induce firms producing them to adapt their plants to use a lower cobalt content or to substitute other materials. The resultant temporary decrease in the production of magnets used by the electrical products industry might require it in turn to reduce output or close plants for a time. The result could be a chain reaction of output reductions and plant closures for each industry dependent on another for products containing the disrupted material. A stockpile release in such circumstances would yield additional social benefits by avoiding or mitigating these adjustment costs.

The following example gives an idea of the loss-benefit interaction arising from the operation of an economic stockpile. Suppose that a disruption reduces annual imports from 2 million to 1 million tons, raising a mineral's world price from $100 to $500 per ton. (It is assumed that the mineral is not produced domestically.) There would be a social loss of $300 million from the increase in the aggregate cost of importing the mineral. (Before the import disruption 2 million tons were imported at a cost of $200 million; after it 1 million tons were imported for $500 million.) There would be an additional loss to consumers of, say, $100 million, from having to reduce consumption by 1 million tons.[21] Finally, the adjustment cost to the economy caused by the higher price might well be $200 million over the six-month period. All this comes to an aggregate social loss of $600 million. A release of 1 million tons would offset import disruption and would, therefore, provide a social benefit of $600 million, but if there were only a 20 percent probability of this import disruption over the decade during which the reserve was held, the probability-adjusted social benefit would be only $120 million (20 percent of $600 million). A stockpile of 1 million tons acquired at $100 per ton and held for ten years (since it is not known when or if supply will be disturbed) is likely to cost $150 million, including interest on the capital invested as well as maintenance. Therefore the release must be less than 1 million tons, for social benefit-cost accounting mandates that the cost of maintaining a stockpile must be less than the probability-adjusted social benefit it provides. In this example, the planned stockpile release would need to be adjusted downward to the point where the social cost of the stockpile was lower than the probability-adjusted social benefit.[22]

Determining the Probability Coefficients for Import Disruptions

How are the probability of reductions in imports of a mineral and the probable duration of those reductions calculated? Expert knowledge of each import source and the application of probability analysis are required. Assuming, for example, a series of estimates such as a 50 percent reduction in imports lasting for six months has a 30 percent probability of occurrence over a ten-year period, and a 25 percent reduction lasting for one year has a 40 percent probability over the same time, it is possible to calculate a combined probability coefficient for a given mineral's average import reduction for an average duration over a ten-year period.[23] It is then possible to multiply the social benefits (or the social costs avoided) by the combined probability coefficient. The results of this calculation give the probability-adjusted social benefit for the entire decade, so that annual benefits are one-tenth the calculated amount. This calculation is made for each mineral chosen for inclusion in the stockpile on the basis of its potential vulnerability to foreign supply disruption.

Since social benefits and costs are incurred over different time spans, comparison between them must be made by comparing the present values of the streams of benefits and costs over time. Annual probability-adjusted social benefits are discounted at the "social rate of discount" or the rate of discount used for determining the present values of benefits and costs of public projects. This rate may be the average real rate of interest on riskless securities such as government bonds, or a rate may be chosen for each public project that reflects the risk or uncertainty of the net benefits from the project.[24]

The Social Cost of Stockpile Maintenance

Holding large stocks in reserve involves storage costs, interest on the capital tied up in inventory, expenses for rotating and upgrading the mineral, and deterioration of its quality. Since minerals purchased are a capital asset that will be sold either while the stockpile is held or at the termination of the program, the inventory purchase price does not count as part of the stockpile's maintenance cost. The inventory is likely to be worth more when sold than it was when purchased. (This will not be true if the inventory is sold at the program's termination and the mineral now costs less than when originally bought.)

Maintenance cost must be estimated from the time an inventory is acquired through the entire anticipated period of its holding. Because the purchase of materials will require several years to meet stockpile goals, annual costs will rise as more units are acquired. These annual costs must be cumulated for the entire period the stockpile is to be held. But wouldn't releases triggered by import disruption reduce maintenance costs? Yes, but only temporarily since inventory should be restocked quickly to deal with possible future difficulties.

Estimates of Socially Optimum Stockpile Levels

The "socially optimum" inventory level for a particular mineral is that which maximizes the difference between the present value of the probability-adjusted social benefits and the present value of the social costs involved.[25] Put differently, the stockpile optimum level is that for which the additional social cost of holding the last unit of a material is just equal to the probability-adjusted social benefit yielded by that unit. A great deal of estimation and analysis is required to calculate an optimum level. The process involves educated guesses about disruption probabilities for various foreign suppliers, estimates of how disruptions would affect world prices, estimates of social losses were there no stockpile releases, and computation of the costs incurred by maintenance of various stockpile levels. Once these and other calculations have been made, final determination of the optimum stockpile level requires a highly complex mathematical model capable of handling a large number of variables and interrelationships.[26]

But is this method reliable, involving as it does so many uncertainties and probability estimates for events that might affect mineral supplies in various parts of the world? No, this procedure is not a surefire panacea. But simply pulling numbers from a hat cannot provide better results and would probably yield still worse. Limited knowledge should not preclude a rational and systematic approach. Moreover, rational procedure based on the principle of social benefit-cost accounting reveals the diverse elements of the problem and the kinds of information, no matter how inaccessible, needed to solve it. Very often a rational approach will prevent gross errors, even though it leads to a result that is considerably less than optimum.

Only a few estimates of socially optimum stockpile levels for particular minerals have been published. Federal agencies have made virtually no attempt to make such calculations, since a government economic stockpile has not yet been seriously considered, but Charles River Associates (CRA) has been commissioned by the government to assess American vulnerability to foreign supply disruption for several minerals, and some of the CRA studies include estimates of socially optimum (or "efficient") stockpile levels.

To cite a few examples, CRA published estimates for chromium based on alternative degrees of supply disruption and optimistic and pessimistic probabilities of occurrence for each disruptive scenario. In the less severe disruption, involving a 15 percent decrease in imports of chromite and chromium ferroalloys, optimal stockpile levels are calculated as equivalent to 4.2 and 6.8 months of consumption for the optimistic and pessimistic scenarios, respectively. The more severe case assumes a 26 percent drop in the same imports. In this instance the socially optimum stockpiles range from 9 to 14 months' consumption for the respective optimistic and pessimistic scenarios.[27] During the 1979–81 period, U.S. chromium consumption (metal equivalent) averaged about 540,000 tons per year, but private inventories were equal to 5 months' consumption, so that the estimated amount of the stockpile would be reduced by this amount to be at a socially optimum level.

The same CRA study estimates socially optimum stockpile inventories for cobalt based on different periods of terminated cobalt supply from Zaire, and of varying possibilities for these contingencies. CRA calculates the optimal stockpile level as ranging from eight months' consumption for the least pessimistic to twenty months' consumption for the most pessimistic; the most likely scenario estimates seventeen months' consumption.[28] (U.S. cobalt use has been declining since the CRA estimates,[29] partly because of the sharp rise in cobalt prices in 1978. Some elasticity via substitution is thus in evidence.)

World manganese sources are more diversified than those for chromium and cobalt, which tends to reduce any particular disruption's impact on global supply. CRA optimal stockpile estimates range from two-and-one-half months' consumption (most optimistic scenario) to over nine months (pessimistic scenario), to as high as sixteen months' normal consumption ("very pessimistic" scenario).[30]

CRA's levels for the platinum-group metals (platinum and palladium) extend from two months' to twelve months' consumption; the estimates for bauxite range from six to twelve months.[31] (The higher stockpile figure for bauxite assumes the existence of a strong bauxite cartel, the likelihood of which has greatly declined in recent years.)

The principal conclusion emerging from this review of CRA's estimates is that economic stockpiles are likely to be cost-effective for only a few commodities. In such cases socially optimum stockpiles (above normal private inventories) would not exceed eighteen months' normal U.S. consumption; the usual level would equal less than a year's need.

Objections to Government Economic Stockpiles

Critics oppose the establishment of the kind of economic stockpile described above on several grounds. One objection is that protection against import disruption in peacetime should be left to the private sector. It is true that industrial users of mineral products hold them in stock to hedge against sharp price increases. In addition, they may have foreign or domestic contracts for future delivery, but the level of private inventories is unlikely to approach the socially optimum level.[32] Private firms adjust their holdings to maximize net profits, while the government should act to benefit all consumers by maximizing net social benefits. Nevertheless, as noted later in this chapter, it may be possible for government to subsidize private inventories until they approximate the socially optimum level.

Another criticism frequently advanced by the mining industry is that a government economic stockpile might discourage domestic production to offset lessened mineral imports. Higher prices for important minerals would of course make it more attractive to mine and process them in the United States. But if a government stockpile were used for anti-inflationary and budgetary purposes, as the NDS has occasionally been, the new incentives for domestic production would disappear. This is a legitimate concern, but not valid grounds for rejecting a program that established rigorous guidelines for releases. The latter should be made only when major import disruptions, originating outside the open market, generated a multiple rise in the world price for a stockpiled material.

A third objection pertains to the sale of stockpile releases on the open market. It is argued that they should be restricted to the most essential uses, with prices no higher than the cost of acquiring the material plus the cost of retaining them in the stockpile. But such a system would impose upon government the responsibility of determining which civilian industries and classes of consumers should be so favored. Not only would this be a dubious role for government in a free enterprise economy, it would constitute a political and administrative nightmare.

Finally, some experts favor controls on exports and on speculative hoarding if stockpile releases are sold domestically. It is, however, administratively very difficult to maintain a domestic market price that differs significantly from the world market, and peacetime export controls have never worked very well.[33] Moreover, the United States has an interest in minimizing the impact of supply disruptions on its major industrial trading partners. As chapter 6 made clear, such countries should be encouraged to maintain economic stockpiles, coordinated with their American counterparts. Hoarding increased domestic inventories after stockpile sales might be limited—perhaps by imposing special taxes. But speculation, after all, may *lower* prices if those holding the material liquidate their stocks in anticipation of an end to supply disruption. Therefore, attempts to control speculation would probably be neither effective nor very useful.

Both government and the general public believe for the most part that any sudden reduction of mineral supplies, whether from domestic or foreign sources, should be dealt with by some form of rationing and price controls rather than by continuing distribution through the market system. This attitude came into play when the government allocated petroleum supplies during the Arab oil embargo of 1973–74. The allocation system not only brought about long lines at the gasoline pumps with enormous consumer frustration and loss of time, it also impaired adjustment of supply and demand. By contrast, the principal method established to distribute releases from the Strategic Petroleum Reserve (SPR) is sale at competitive prices. Moreover, the competitive price rule would also apply to SPR oil sold abroad in order to meet U.S. obligations under the International Energy Program.[34]

Alternatives to a Government Economic Stockpile

Several plans have been suggested as preferable to a government economic stockpile. Only one of these alternatives, encouraging a substantial increase in private inventories of nonfuel minerals subject to a high probability of import disruption, might accomplish the same objectives more economically than a government economic stockpile. One alternative—subsidization of domestic mineral production—would neither be cost-effective nor achieve economic stockpile objectives. Still other measures, such as R&D to develop substitutes and to conserve nonfuel minerals and promoting foreign production in more secure areas, might reduce vulnerability, but would not achieve the same objectives as an economic stockpile.

Subsidizing Domestic Production of Minerals

There are several nonfuel minerals for which the United States depends on foreign sources, which could be produced from lower-grade domestic reserves, but at a considerably higher cost. They include chromium, cobalt, manganese, and platinum. A recent study of the U.S. resource base and potential output for these four minerals concludes that without government subsidies only small amounts of American platinum are likely to be mined between 1980 and 1995.[35] Under free market conditions, some cobalt may be produced by 2010, depending upon its price, but domestic chromite and manganese are unlikely candidates without significant government inducement (i.e., government purchases at prices several times the world level).

Even so, domestic manganese production is unlikely under any condition, except possibly as a secondary material from other mining operations. In no case would domestic output amount to more than a fraction of U.S. requirements. Furthermore, unless new reserves of these minerals are discovered, the resource base would be exhausted within a few years.

Congress, the Reagan administration, and public interest groups are divided over the desirability of special incentives to develop domestic minerals that could not otherwise compete with imports. In 1982 and 1983 several bills were introduced in Congress to fund government purchases under Title III of the Defense Production Act of 1950. Supporters who testified included Department of Defense officials and representatives of the American Mining Congress.[36] Since the government would contract to purchase minerals for more than producers might expect in the market, the program would subsidize domestic production. The contribution of domestic production to national defense is emphasized, but the strong support by mining interests—particularly those companies with claims on areas known to have deposits of such minerals—suggests that the program is also being promoted for the special benefit of the industry.[37]

Some administration officials opposed subsidies for domestic production of low-grade minerals. For example, David Stockman, the former Office of Management and Budget (OMB) director, argued that the United States should let market forces improve the competitiveness of American industries and that national security objectives should be met by purchases for the strategic stockpile at world market prices.

In testimony before the House Subcommittee on Economic Stabilization about the Defense Industrial Base Revitalization Act (H.R. 5540),[38] Robert Wilson, director of the Department of Commerce's Office of Strategic Resources, stated that the "Administration opposes provisions of the Bill which raise authorizations for Title III to $1 billion a year for five years. We do not plan to make use of Title III authorities to subsidize domestic industry. Instead, the Administration intends to rely on the marketplace to improve competitiveness of our industries and help reduce dependence on foreign sources of critical materials."[39] Wilson went on to say that "proposed programs under Title III would constitute a major intervention in commodity markets and could represent a substantial drain on scarce budgetary resources, and may contribute to a worldwide overcapacity."[40] Dr. John D. Morgan of the Bureau of Mines also testified against the proposal. The secretary of defense favored government subsidization of domestic production, but proposed funding the programs from his department's budget.[41]

A U.S. General Accounting Office (GAO) report raised serious questions about the economic efficiency of acquiring cobalt for the stockpile from domestic sources at $25 to $30 per pound when cobalt could be purchased for

$15 per pound or less from Zaire.[42] The GAO also suggested that "federal incentives to foster domestic mining of limited reserves [of cobalt] could result in the unwarranted depletion of domestic supplies that might better be saved for future use." Furthermore, stockpile acquisitions gained by exploiting "marginally economic reserves of cobalt may not only be obtained at great economic cost, but at great environmental cost as well."[43]

Some favor subsidized domestic production on the grounds that an additional ton of domestic output each month is equivalent to an inventory of six tons that would be released over a six-month period. Releases of stockpile, however, are unlikely more than once or twice in a decade, while domestic output equivalent to monthly stockpile releases would be continuous. Should foreign supply be greatly diminished, stockpile releases equal to a large part of normal domestic consumption could be required to offset a world price rise and a decrease in availability. Domestic production could not provide the increased amount of the mineral required over a short period of foreign supply interruption.

The additional cost per ton (over world market price) for domestic production should be compared with the cost per ton to stockpile that material. If the acquisition cost at the world price is $1,000 per ton and the annual cost of holding the inventory is 15 percent of the acquisition cost (including interest), the stockpile cost is $150 per ton annually. If additional domestic output costs more than 15 percent above the world price, maintaining a stockpile would be cheaper than subsidization. The fact that the material would come from domestic rather than foreign sources would make no difference in terms of social costs: a dollar spent for domestic production has the same social cost as a dollar spent for imports. Hence, a stockpile is likely to be more economical than subsidized domestic production of a mineral unless its domestic cost is not much above the world price. In 1984 it would have required a subsidy of more than twice the free market price to produce cobalt domestically.

A recent simulation study by a government agency compared net social benefits (social benefits less social costs) of a program to subsidize domestic cobalt production with a program to increase private inventories obtained abroad. The prognosis under a worst-case disruption scenario (an interruption of cobalt imports from both Zaire and Zambia) was that the private inventory enhancement program would yield positive net social benefits, whereas the program to promote domestic production by subsidy would produce negative net social benefits. Only in the highly unlikely case of a long period of severe reduction in cobalt imports was domestic production found to offer positive net social benefits. This was true in part because it would require several years to create capacity for significant domestic production. Short-term supply disruptions were regarded as relatively probable, but severe disruptions lasting more than two years were seen as very improbable.[44]

Promoting Diversification of Foreign Supply

Under the authority of the Defense Production Act of 1950, the government has encouraged development of new sources in areas believed unlikely to interrupt supply. Several techniques are possible for this endeavor, including long-term contracts for output, U.S. government loans, and Overseas Private Investment Corporation (OPIC) investment guarantees. In terms of supply vulnerability, social benefits would come, if at all, mainly from the effect additional supply had on world prices when more vulnerable areas reduced output. The United States cannot count on materials from other countries being made available exclusively to it or at predisruption prices, even if foreign facilities are owned by American firms. Governments of countries where investments are made are likely to insist that their mineral products be sold in the most profitable markets. This should be kept in mind when weighing the social benefits that could result from promotion of mineral capacity abroad. Nevertheless, U.S. government incentives to expand, say, Mexico's manganese capacity or chromium production in the Philippines ought to be given serious consideration. The fact that neither of these foreign sources can be regarded as "secure" from internal conflict that would jeopardize exports does not invalidate this recommendation. Civil disturbances in these countries are unlikely to occur at the same time that supply from other sources is disrupted.

R&D for Substitution and Conservation

Given enough time and adequate price incentives, many if not all of a mineral's industrial uses can be met by substitutes. Also, alternative processes that greatly reduce needs for a mineral may be employed without sacrificing performance, but costly R&D is often necessary to develop such substitute materials or develop conservatory processes. Private firms are unlikely to embark on such R&D so long as ample supplies of a material are available at prices not much higher than substitutes. The social benefits derived from having these alternative materials or processes available in the event of supply disruption could be large enough to justify government subsidies for their development. The prospective gain might even warrant their use on a limited scale so that they could readily be adopted on a larger scale in appropriate circumstances.

The most important use of cobalt is for the aerospace industry's superalloys, and a few years ago it was regarded as irreplaceable in jet aircraft engines. However, cobalt's sharp price rises in 1978 and 1979 (following an invasion of Zaire) led to R&D programs to reduce its use. For example, a frequently used alloy in jet blades contains over 50 percent cobalt. Three alternatives containing between 18 and zero percent cobalt have been developed and the latter is reported to have operating properties superior to the currently used cobalt alloy. An upgraded version of the F-15 and F-16 military aircraft engine uses a nickel-aluminum-molybdenum alloy containing no cobalt.[45] Machine tools and construction machinery also utilize cobalt, but here again new alloys for high-speed steel tools are expected to replace most cobalt-bearing alloys.[46]

During the past decade research has been carried on by the Bureau of Mines in cooperation with aluminum consortia to produce alumina from nonbauxite sources, including domestically abundant clay and shale. In 1980 a plant was designed for clay resources.[47] Some nonbauxite sources require no more energy for production than bauxite does,[48] but estimates of operating and capital costs indicate that production from nonbauxite sources is somewhat more expensive. After attempts to form a bauxite cartel failed and Australia became the leading producer, with the world's second-largest bauxite resources, the U.S. aluminum industry evidenced little interest in substitutes. For these reasons the BOM is terminating most of its research on nonbauxite sources.[49]

Subsidizing Private Inventories

Private inventories of materials subject to import disruption do not usually attain socially optimum levels, but the accumulation of additional amounts could be subsidized by the government. If firms using a material susceptible to foreign supply disruption normally maintained three months' inventories (whether in raw or semiprocessed form), they might be given subsidies to stock enough for six months or a year. It would be unnecessary to underwrite the full additional cost since the firms would benefit from their larger inventories should a foreign supply cutback result in a big price rise. The subsidy program would need to avoid an expansion of inventories so rapid as to affect world prices greatly. Subsidization could be through low-interest government loans that would fall due when inventories declined below target amounts. Furthermore, those who held excess inventories could be protected, should they go out of business or reduce their output, by permission to repay government loans with inventory materials valued at the original purchase price.

There may be objections because under this scheme government would subsidize private benefits in case of sharp world price rises. True, but there are social benefits to consider as well: the higher inventory accumulations would reduce transfers abroad and other social losses. If the firms accumulating excess inventories assumed the cost of their maintenance and paid some interest on the government loan as well, both parties would share the cost of inventories that provided both private and public benefits.

A private inventory inducement program could be integrated with the existing national strategic and critical stockpile, which releases materials to defense industries in the event of major foreign supply disruption. To the extent that a firm supplied products for defense, it would not need to participate in the inventory inducement scheme. It would obtain its requirements from strategic stockpile releases, thus reducing the cost of inventory inducement.

There are several advantages to having private firms stock the extra materials rather than requiring government to do so. Privately held inventories would take the form specifically required in the production process (e.g., ferromanganese rather than manganese metal). These forms could readily be changed to accord with new technologies and product demands. And they would be located near where they were used, not perhaps thousands of miles away in a government stockpile. Unlike the present government stockpile, whose materials deteriorate without special attention, private inventories could continually be revolved by drawing down older materials and replacing them with newer ones in the ordinary course of business.

Conclusions

America's strategic stockpile program is oriented to a three-year all-out war scenario, but not to peacetime conditions, except possibly for foreign materials needed for defense production. A conventional war lasting several years during which the U.S. production system would not be severely damaged seems unlikely. More probable are a devastating strategic nuclear war or internal conflicts and limited regional wars in areas that supply nonfuel minerals to the Western industrial countries. Yet the United States has virtually no government programs to protect the civilian economy against such eventualities.

Although I do not favor termination of the U.S. strategic stockpile program per se, I do think the assumptions on which it is based should be carefully examined by a panel including experts drawn from outside the minerals industry and the executive branch.[50] Such a review might well indicate that the size and mineral composition of the NDS should be altered and its size cut back. In fact, after a two-year study by the National Security Council (NSC), the Reagan administration proposed in mid-1985 a major revision of the NDS program to Congress.[51] By all accounts the current program is poorly designed to meet its basic goals.

Excluding the case of a major war involving the United States, the American economy's vulnerability to foreign supply disruption of essential nonfuel minerals arises from the possibility of: (1) a foreign cartel able to restrict exports so as to raise world prices; (2) a politically motivated embargo similar to the 1973–74 Arab oil embargo; and (3) civil unrest or regional conflict in one or more major producing countries that would curtail mineral production or transportation. Given the nature of mineral markets, the diversification of sources, and the political relationships among countries supplying those minerals the industrial countries depend on, neither of the first two contingencies is likely. Limited wars, civil unrest, and internal revolution are the most probable sources of dangerous foreign supply interruption.

The main damage to the domestic economy in peacetime would arise from an increase in a mineral's world price. Foreign mineral supply disruption is likely to damage the American economy severely only if the United States and other major industrial countries depend on countries or regions with great political instability for more than 25 percent of their requirements; if there are no readily available substitutes for most uses; and if sources outside developed nations are concentrated in a few currently or potentially unstable countries.

The U.S. economy is subject to some import disruption for a number of commodities. But only a handful of nonfuel minerals justify costly government action to avoid or mitigate the consequences of supply disruption. Various private and government-sponsored reports differ as to which minerals should be included in this group, but most frequently mentioned are bauxite-alumina, chromium, cobalt, manganese, and the platinum-group metals. The materials warranting special government action should be chosen by social benefit-cost accounting. This method requires detailed analysis of social costs for various degrees of supply disruption; it also means that probabilities of occurrence must be calculated.

The discussion in this chapter has largely focused on the impact import disruption might have on the American economy, but U.S. concern should encompass consumers in the entire non-Communist world for two reasons. First, a supply disruption in any major source of a mineral will increase its world price. That would mean a social cost for the U.S. economy even if the bulk of imports came from an unrestricted source like Canada. Second, if production in Western Europe or Japan is reduced because of a multiple rise in the price of a nonfuel mineral, that will inevitably affect the U.S. civilian economy and may also impair the defense efforts of its allies.

A government economic stockpile has been suggested to counteract possible social costs arising from foreign supply disruptions in peacetime. Stockpile releases should be sold at world market prices; there should be no rationing of domestic supplies in a peacetime economy. The socially optimum level of each mineral inventory should be determined on the basis of social benefit-cost accounting, with potential social benefits adjusted for the estimated probabilities of a range of scenarios. Although this study makes no attempt to determine all the minerals needed for an economic stockpile, or to estimate the socially optimum level of any stockpiled mineral, a review of existing studies on the subject suggests that the number of minerals for which an economic stockpile would be socially efficient is likely to be small, and that optimum inventories would be less than eighteen months' consumption.

A government program to subsidize additional private inventories of critical minerals up to the estimated socially optimum level could serve the same purposes as a federal economic stockpile. Subsidies could take the form of low-interest loans for additional inventory accumulation, with these loans coming due when inventories decline below targets. Privately held inventories would have the advantage of providing the mineral in forms specifically required for production, and these forms could be readily adjusted to changing technologies and product demand. Moreover, they would be located where used, not thousands of miles away in a government stockpile. Also, materials held in private inventories could be continually revolved rather than allowed to deteriorate like the national strategic stockpile. The cost of maintaining above-normal inventories should be shared by government and private holders since the excess would provide both private and social benefits when and if used.

Government subsidies for domestic production from low-grade deposits would be more costly and less efficient than stockpiling for dealing with foreign supply disruptions. On the other hand, subsidies to explore and develop technologies that would eliminate or conserve on the industrial use of vital imported minerals may be justified as a kind of insurance against long-lasting supply disruptions. Government promotion of R&D would not obviate the need for an economic stockpile or enhanced private inventory accumulation, since it would take some time to produce and install facilities based on the new technology developed. In some cases at least the new methods would not be economically attractive as long as imported minerals remained fairly cheap.

Appendix 7-1

Commodities Held by National Defense Stockpile

TABLE A7-1. National Defense Stockpile Inventory of Strategic and Critical Materials, March 31, 1983

Commodity	Unit	Goal	Inventory
Aluminum (metal)	st Al metal	7,150,000	3,813,679
Aluminum oxide (abrasive grain)	st ab grain	638,000	259,124
Antimony	st	36,000	40,402
Asbestos (amosite)	st	17,000	42,534
Asbestos (chrysotile)	st	3,000	9,957
Bauxite (refractory)	lct	1,400,000	199,926
Beryllium (metal)	st Be metal	1,220	1,061
Bismuth	lb	2,200,000	2,081,298
Cadium	lb	11,700,000	6,328,809
Chromium (chemical and metallurgical)	st Cr metal	1,353,000	1,324,923
Chromite (refractory)	sdt	850,000	391,414
Cobalt	lb Co	85,400,000	45,995,714
Columbium group	lb Cb metal	4,850,000	2,532,419
Copper	st	1,000,000	29,048
Cordage fibers (abaca)	lb	155,000,000	0
Cordage fibers (sisal)	lb	60,000,000	0
Diamonds (industrial)	kt	29,700,000	38,723,355
Fluorspar (acid grade)	sdt	1,400,000	895,983
Fluorspar (metallurgical grade)	sdt	1,700,000	411,738
Graphite (natural, Ceylon, amorphous lump)	st	6,300	5,499
Graphite (natural, Malagasy, crystalline)	st	20,000	17,899
Graphite (natural, other sources)	st	2,800	2,804
Iodine	lb	5,800,000	7,525,930
Jewel bearings	pc	120,000,000	71,246,385
Lead	st	1,100,000	601,032
Manganese (dioxide, battery grade)	sdt	87,000	218,405
Manganese (chemical and metallurgical)	st Mn metal	1,500,000	1,970,722
Mercury	fl	10,500	182,815
Mica (muscovite block, stained and better)	lb	6,200,000	5,212,445
Mica (muscovite film, 1st and 2d qualities)	lb	90,000	1,252,138
Mica (muscovite splittings)	lb	12,630,000	18,157,850
Mica (phlogopite block)	lb	210,000	130,745
Mica (phlogopite splittings)	lb	930,000	1,678,742
Molybdenum group	lb Mo	0	0
Morphine (sulphate and related analgesics)	ama lb	130,000	71,303
Natural insulation fibers	lb	1,500,000	0
Nickel	st Ni + Co	200,000	32,209
Platinum group (iridium)	tr oz	98,000	23,590

Continued on next page

TABLE A7-1—*Continued*

Commodity	Unit	Goal	Inventory
Platinum group (palladium)	tr oz	3,000,000	1,255,003
Platinum group (platinum)	tr oz	1,310,000	452,642
Pyrethrum	lb	500,000	0
Quartz crystals	lb	600,000	2,063,827
Quinidine	av oz	10,100,000	1,874,504
Quinine	av oz	4,500,000	3,246,164
Ricinoleic/Sebacic acid products	lb	22,000,000	12,524,242
Rubber	mt	864,000	120,475
Rutile	sdt	106,000	39,186
Sapphire and ruby	kt	0	16,305,502
Silicon carbide (crude)	st	29,000	80,550
Silver (fine)	tr oz	0	137,505,946
Talc (steatite block and lump)	st	28	1,081
Tantalum group	lb Ta metal	7,160,000	2,426,387
Thorium nitrate	lb	600,000	7,131,812
Tin	mt	42,700	193,642
Titanium sponge	st	195,000	32,331
Tungsten group	lb W metal	50,666,000	79,181,354
Vanadium group	st V metal	8,700	541
Vegetable tannin extract (chestnut)	lt	5,000	15,068
Vegetable tannin extract (quebracho)	lt	28,000	135,506
Vegetable tannin extract (wattle)	lt	15,000	15,386
Zinc	st	1,425,000	378,316

Source: Federal Emergency Management Agency, *Stockpile Report to the Congress, October 1982–March 1983* (Washington, D.C.: U.S. Government Printing Office, 1984), table 2.

Notes: ama lb = pounds of anhydrous morphine alkaloid; av oz = avoirdupois ounce; fl = flask (76-pound); kt = carat; lb = pound; lb Cb = pounds of contained columbium; lb Co = pounds of contained cobalt; lb Mo = pounds of contained molybdenum; lb Ta = pounds of contained tantalum; lb W = pounds of contained tungsten; lct = long calcined ton; ldt = long dry ton; lt = long ton; mt = metric ton; pc = piece; sdt = short dry ton; st = short ton; st Ni + Co = short tons of contained nickel plus cobalt; st V = short tons of contained vanadium; tr oz = troy ounces

Appendix 7-2

Analysis of a Cartel's Ability to Raise the Price of a Material and Reap Monopoly Profits by Restricting Output

The elasticity of world demand for a cartel's output is lower (and hence the cartel's monopoly power is greater), the smaller the elasticity of world demand, the larger the cartel's share of world output, and the smaller the supply elasticity of noncartel sources. Elasticity of world demand for a material depends upon the elasticity of demand for products that need the material in their fabrication, and upon the proportion of the final product's cost represented by the material. If the elasticity of demand for the final product is −2.0, but the material constitutes only 10 percent of its value, the elasticity of demand for the material will be only −0.2. If the short-run elasticity of supply from sources outside the cartel is zero, and the cartel contributes 50 percent of the total world supply of the material, elasticity of demand for the cartel's output would be −0.4. Thus the cartel could raise the price 25 percent by reducing output 10 percent, thereby increasing total receipts and monopoly profits. If, however, the elasticity of supply from noncartel sources were −0.6, the elasticity of demand for output of the cartel would be −1.0, and the cartel could not increase its total receipts by restricting output. Under these conditions the cartel would have little incentive to restrict output or sales in order to raise world prices.

Appendix 7-3

Vulnerability to Import Disruption of Five Nonfuel Minerals

Bauxite

A number of foreign countries produce bauxite. Only Australia (31 percent) accounts for more than 15 percent of total output (see table A3-2). Moreover, the principal producers are economically and politically widely diversified, and lack the political cohesion to form a strong cartel or enforce an embargo. Ordinarily, therefore, bauxite would not be regarded as highly susceptible to foreign supply disruption, despite its great importance to the world economy and the absence of short-run substitutes. Nevertheless, considerable concern has been expressed over the world economy's vulnerability to cartel action in bauxite,[52] mainly because of the International Bauxite Agreement established in the wake of OPEC's price increases. Eleven IBA members control about two-thirds of world bauxite output and more than one-third that of alumina.[53]

A major purpose of the IBA was to increase members' export revenues by imposing new taxes on bauxite exports. Jamaica took the lead with a production levy equal to 7.5 percent of the average realized price of primary aluminum ingots divided by 4.3—roughly the number of pounds of bauxite required for a pound of aluminum. Some other IBA members increased their taxes on bauxite production, but not by as much as Jamaica. The IBA, while far from being a cartel, was successful in raising prices by an average of $10 per mt in 1974. Each year the IBA recommends a price for a reference grade of bauxite and a price for alumina, both based on aluminum metal prices.[54] Although bauxite prices did rise during the remainder of the 1970s, in 1981 and 1982, average prices (in constant 1981 dollars) were lower than in 1974, and there was a substantial decline in 1983. Some large producers, including Australia, have paid little attention to IBA price recommendations.[55] Jamaica, which had raised its export taxes more than any other bauxite-producing country, lost some of its world market share and subsequently reduced its levies.

The formation of a strong bauxite cartel is constrained by the fact that a large proportion of bauxite comes from integrated international aluminum firms, each of which has investments in bauxite production in several countries. The companies can skew production and investment in additional capacity to those areas where conditions are most favorable. Moreover, there is a long-term limit on price rises, since a doubling or tripling would encourage aluminum production from nonbauxite sources that the United States and other developed countries have in almost unlimited quantity. So a bauxite cartel to raise the world price of aluminum substantially does not seem likely to be formed.[56]

Supply disruption from internal strife or revolution is likely in one of the major bauxite-producing countries, particularly Jamaica, Guyana, Guinea, or Suriname. However, this is unlikely to occur in more than one country at a time, so that world prices would not be profoundly affected. In any case, supplies would probably not be disrupted for more than a few months. It should be mentioned, however, that a production disturbance in Jamaica caused by civil disorder would seriously concern the United States over the short run, because 80 percent of American "mixed" bauxite requirements come from Jamaica.[57] Jamaican alumina exports could also be cut off by internal strife, but other sources, well diversified among a number of countries, could take up the slack.

Chromium

The situation with respect to chromium is different from that of bauxite. First, known commercial quantities of chromium are found in only a few countries, and non-Communist output is heavily concentrated (73 percent in 1984) in South Africa, Turkey, and Zimbabwe. (Small amounts are produced in the Philippines, Brazil, India, Finland, and Yugoslavia.) Second, no substitutes exist for certain uses of chromium: high-strength steels, high-temperature metals, and corrosion-resistant alloys essential for making petrochemicals and for the manufacture of jet engines and power plant equipment.[58]

Because chromium makes up only a small percentage of the value of end products and because there are no readily available substitutes for many of its uses, demand for it is highly inelastic. Without stockpile releases a foreign supply disruption would dramatically raise prices. There are, of course, substitutes for the chromium used in decorative stainless steels, automobile trim, flatware, refractories, and some chemicals. A 1978 study concluded that U.S. chromium consumption could be reduced by up to one-third within five years by using available technology to substitute alternative materials or processes and to recover and recycle waste chromium. The same study points out that a research program could reduce consumption by an additional one-third within ten years. However, even if all these measures were taken, unsubstitutable uses are expected to require at least 180,000 st per year (approximately one-third of 1980 consumption) for the foreseeable future.[59]

Most analysts see the main potential problem as civil strife in South Africa and Zimbabwe. Formal cartel action that would involve the collaboration of South Africa with the USSR, Zimbabwe, and Turkey is politically unlikely, although informal collaboration with South African efforts to raise prices is a possibility (though the present South African government would probably hesitate to incur Western hostility by such an action). A politically motivated embargo by South Africa and other major chromium exporters is even more unlikely. A more likely, but still remote, possibility is a UN embargo on South Africa with the participation of the United States and other non-Communist industrial powers. The UN embargo against Rhodesia (now Zimbabwe) was not very effective, and the industrialized democracies are simply too dependent on South African minerals to shoulder the cost of an embargo, even if they were politically so inclined.[60]

In 1984 the largest chromite producers were the USSR and South Africa, each with 32 percent of world output, followed by Albania (9 percent); Turkey (6 percent); and Zimbabwe (5 percent). Most of the USSR output is used domestically or shipped to other members of the Soviet bloc. About 84 percent of the world's reserves of chromite are held by South Africa, followed by 11 percent by Zimbabwe, and only 2 percent by the USSR. The non-Communist world is therefore likely to depend heavily on South Africa and Zimbabwe for supplies of chromite in the future. However, other non-Communist countries accounted for about 22 percent of world output in 1984. Even if South African exports of chromite were temporarily disrupted, a substantial amount of it would be available from reasonably secure sources of supply.

Cobalt

About 40 percent of U.S. cobalt consumption goes for "superalloys" used mainly in industrial and aircraft gas-turbine engines; 15 percent is used for magnetic materials; 12 percent for catalysts; 12 percent for driers; 10 percent for metal-cutting and mining tool bits; and 11 percent for other purposes. Approximately half of the 15.3 million pounds consumed by the United States in 1980 was devoted to essential uses that neither substitution nor conservation could replace.[61]

The United States has mined no cobalt since 1971. Except for cobalt recycled from scrap (about 10 percent of domestic consumption), America is wholly dependent on imports. Domestic reserves of low-grade ores could become commercially feasible sources only if import prices rose substantially (say, by 50 percent) for a long time. Within several years the resulting domestic production could equal 20 percent or more of projected U.S. consumption. Zaire and Zambia mine over 60 percent of total world cobalt output and over 70 percent of non-Communist output. The same two countries account for over half of total world reserves.

Because cobalt, like chromium and other alloy metals, constitutes only a small fraction of end-product cost, short-run demand tends to be inelastic. Prior to 1978 a Zairian-Belgian selling syndicate maintained effective control over cobalt's world price, but since then the syndicate's power has ebbed. Even if it should regain its previous market control, adoption of a short-term maximum price is not likely, because that would stimulate expanded production elsewhere (including the United States): over time higher prices could lead to a reduction in demand through substitution and conservation.[62] There also appears to be little basis for a politically oriented embargo by Zaire and Zambia, an action that might also entail the loss of a large portion of their market. The most likely source of supply disruption from one or both central African producers is civil unrest in these politically unstable countries, or another invasion like the 1977 incursion into Zaire from Angola.

Manganese

Manganese is mainly used as ferromanganese (an alloy of iron and manganese) to produce steel. There is no satisfactory substitute. Manganese is also used in nonferrous alloys with copper and aluminum to manufacture dry-cell batteries and a variety of chemical products.

The United States has no mine production of manganese and no commercial reserves at present prices. Half of America's manganese imports are in the form of ferromanganese, mainly from South Africa and France. Recycling is not significant. More than one-third of the world's manganese is mined in the USSR. South Africa produces 50 percent of the non-Communist world's manganese; most of the rest comes from Brazil, Australia, Gabon, and India (in that order).

U.S. ferromanganese production has been declining even as imports have been rising. For example, in 1969 the United States supplied nearly three-fourths of its own ferromanganese requirements, but by 1979 the figure was down to less than 30 percent; imports supplied the remainder.[63] However, the potential vulnerability problem in peacetime is with manganese ore, not the ferromanganese alloy. Thus most of the U.S. ferromanganese supply comes either from South Africa or from other industrial powers that are themselves dependent upon the few countries that produce the bulk of manganese ore.

The short-run demand for manganese is highly inelastic, but high manganese prices would eventually result in conservation through investment in external desulphurization and continuous casting processes. U.S. private stocks of manganese ore and ferromanganese at the end of 1981 were 1,035 st, or nearly 100 percent of consumption for that year. Manganese ore and ferromanganese in the national strategic stockpile were about three times the level of private stocks.

As with chromium, the major potential source of manganese supply disruption is civil unrest in South Africa. However, high manganese prices could encourage an expansion of output elsewhere, particularly Australia and Gabon, which have fairly large reserves. A politically motivated embargo on South African manganese exports is exceedingly remote; again, as in the case with chromium, a more likely contingency would be the imposition of UN sanctions. A manganese ore producers' cartel involving South Africa, Australia, Gabon, and India is conceivable, but the diverse political, economic, and strategic interests of the largest producers render such cooperation highly unlikely.[64]

The Platinum-Group Metals

Platinum, palladium, and iridium are the principal platinum-group metals used in industry. In 1979, platinum itself accounted for over 50 percent of U.S. industrial consumption of these metals; palladium made up 40 percent. In recent years the major use of platinum group metals in the United States has been in catalytic converters for automobile exhaust systems, but they are also used in chemicals, petroleum refining, electrical apparatus, and dentistry.

Domestic mine production supplies a negligible portion of U.S. consumption of these metals. The recycling of scrap is, however, a significant source—about 15–20 percent of U.S. consumption. Mining domestic deposits could ultimately provide about 20 percent of platinum and nearly all palladium consumption.[65]

The two major producers are South Africa and the USSR. In 1984, South Africa accounted for about 41 percent of world output, the USSR for 52 percent (most of which, however, was consumed within the Soviet bloc). The only other significant supplier is Canada, with 5 percent of global production in 1984. Canada's output is insensitive to price since it is a by-product of nickel production. The vast bulk of known reserves are in South Africa.

The main sellers of platinum-group metals on the world market are South Africa and the USSR; the latter dominates in palladium and the former in platinum. Although formal collaboration between these two countries in a cartel is highly unlikely, if either reduced sales to raise prices, the other country might support the move.[66] Among the possible causes of platinum and palladium supply disruption, Soviet action has a high probability. In 1975 the USSR nearly doubled palladium prices and would undoubtedly seize any opportunity to increase them further. South Africa is far less likely to withhold supplies in order to raise prices, but the possibility of civil turmoil in that country threatens its ability to continue production of these minerals as well as chromium.

8

Environmental Protection versus Mineral Availability: A False Dilemma

There is a fierce and growing conflict between the U.S. mining industry and environmentalists over air and water pollution, land reclamation, and the exploration and mining of public lands. The struggle is highly politicized: environmentalists seek rigorous enforcement of existing laws and enactment of even more stringent regulations to circumscribe the mining industry's operations, while the industry lobbies for the relaxation of present legal constraints and the defeat of new ones. The mining industry couches its arguments in terms of desirable social goals like economic growth, high employment, an improved balance of payments, and the greater availability of minerals for national defense. Environmentalists contend that preservation of the environment involves social values that should take priority over those associated with the production and processing of minerals. What is needed is a rational approach to this conflict based on the maximization of net social benefits.

Domestic exploration and production promotes the availability of nonfuel minerals in two major ways. It expands world supply, thereby helping to moderate prices at home and worldwide. By reducing dependence on mineral imports, domestic production may contribute to the availability of supplies if they are not available from abroad.[1] The availability of nonfuel minerals to the economy is only one of the social utilities derived from their domestic production. The industry contributes to the national product and to employment. A serious cutback in mineral output (e.g., because of stringent pollution abatement requirements or restrictions on where exploration and mining are allowed) may create severe problems of economic adjustment for the mining sector and for communities and industries economically dependent upon it. Any comparison of the social benefits brought by environmental preservation with the corresponding social costs must therefore include all of them, not simply those directly related to the availability of minerals.

Although both social costs and social benefits of environmental preservation are difficult to measure, wise policy nevertheless justifies making the best possible estimates. Any other course would lead to decisions based solely on conflicting positions of special interest groups. For example, some environmentalists would undoubtedly be willing to abolish the entire copper or chemical industries because they are to some degree hazardous to man and nature: conversely, those who speak for these industries may oppose anything that would lower profitability. But the social costs of either policy would clearly be prohibitive. The problem, then, is to determine just how much environmental damage society ought to accept without destroying entire industries or large portions thereof.

The Economics of Pollution

Some costs of productive activities that defile the environment are borne by members of society who have no part in causing them. These costs are sometimes called external diseconomies and are regarded as social costs of production, as contrasted with its private costs. The question of who should foot this bill has long been debated. Take atmospheric pollution, for example. One school holds that the owner of an industrial complex should pay for the damage suffered by those exposed to airborne emissions from the plant, or that such other actions be taken as taxing the owner an amount equal to the external diseconomies or requiring that the operation be moved to a location where it would injure no one. This view is sometimes called the "polluter pays" principle.[2] The problem of dealing with social cost, however, is not as simple as it first appears. How can the individuals who suffer from smoke or other environmental damage be compensated by an amount equal to the damage they sustain? A tax equal to the estimated social cost could be assessed on the owner of the plant. But since the proceeds are then used for a variety of purposes by the taxing authority, the benefits individuals receive may bear little relation to the harm they actually suffer or suffered. A more satisfactory approach is to require (or induce by emission taxation) the owner to pay for the elimination or reduction of harmful emissions. This approach, too, raises basic questions. First, who ultimately bears the cost of this remedial action? Second, if the paymaster of last resort is in fact society, how much will it be willing to spend for the social benefits of clearer air?

In order to relate these questions directly to the minerals industry, let us assume that the air-polluting agent is a smelter in a copper mining complex. Different methods can reduce sulfur dioxide (SO_2) emissions (the principal form of pollution) by differing amounts; the cost of abatement rises with the degree to which emissions are reduced. Alternatively, the mining firm might be required to replace the existing smelter with one that emits little or no SO_2, but at a considerable capital cost and in some cases an increase in operating costs.[3] Depending upon the nature of the market, the company may add the capital and recurring costs of SO_2 abatement to the price of the product, or it may have to absorb them by reducing profits. If in the latter case profits are already low or nonexistent, the company may be unwilling or unable to accept losses and would shut down. As a result, the local community would lose employment and revenue, resulting in considerable social costs. In order to avoid the losses that a shutdown of the mining complex would bring, the community might agree to subsidize the pollution abatement through local taxation. Thus instead of the polluter, the community itself might pay to reduce pollution in order to avoid the larger cost of losing the mine. Alternatively, of course, if the community had any say in the matter, it might opt to accept the pollution rather than paying extra taxes *or* losing the mine.[4]

If we are dealing with the copper-mining industry as a whole rather than with an individual facility, the social benefits from pollution abatement might be paid for at the social cost of higher copper prices and for finished goods using it, or the social cost might take form as a national reduction in copper output. Such a reduction—a consequence of mine closures and reduced incentives to develop new capacity—would decrease GNP, but copper supply could be maintained by increased imports. (Copper is not significantly vulnerable to foreign supply disruption except under conditions of all-out war, so increased dependence on imports would pose little threat to the availability of copper for the domestic economy.) There would be a negative impact on worldwide copper supply and hence on copper's world price, but the increase in world prices caused by a small U.S. output reduction would be minimal. In this case, the social cost of pollution abatement would mainly be the aggregate reduction of GNP due to lost production. This would need to be compared with the social benefits arising from cleaner air.[5]

Measuring Social Costs of Pollution

Environmental policy is concerned with comparing the social benefits of elminating or reducing the social costs of pollution with the social costs of pollution abatement. Unlike the cost elements for removing pollutants, such as capital equipment, labor, and energy, or the cost in terms of reduced productivity, the social costs of pollution constitute disutilities that are not traded in the market and have no market price. They can, however, be measured indirectly. Many of these disutilities have physical effects such as the destruction of crops or livestock, the deterioration of paint, or physical illness. Monetary values can be placed on crop losses, livestock, medical care, and work time lost due to pollution-related illness.[6] It is even possible to place a monetary value on human life in terms of the capitalized value of wages lost or other criteria.[7]

How much society is willing to pay to reduce the incidence of suffering and loss from illness and premature death caused by pollution may be calculated in various ways, all of which are inadequate. One way is to add up what society's representatives in Congress will appropriate or mandate that industry must pay to reduce harmful environmental factors. But Congress is guided by what costs the public will accept. If it is shown that a 50 percent improvement in air quality would reduce the incidence of serious illness and death by two-thirds, a social cost of $100 billion over several years might be acceptable; but a social cost of 10 percent of the national income ($350 billion) probably would not. Within this range, a democratically elected Congress and administration must decide how much of a reduction in serious illness and death the public is willing to buy, given the social cost for removing pollutants from the air. Thus, in a very crude and often irrational way the political process determines the monetary value of human suffering and premature death. Of course, the more information about these relationships and the better educated the public and its representatives are regarding them, the more rational the decisions are likely to be.

Sometimes critics reject social benefit-cost analysis for environmental issues because the methodology and data are inadequate to make reliable quantitative estimates of both the social benefits and the social costs of environmental protection.[8] Granting the inadequacies of the tools available to social scientists, the alternative is worse: emotional appeals and rhetoric are no substitute for a rational examination of the trade-offs between conflicting social values.

Application of Pollution Abatement Standards

Most environmental economists rely on three basic principles for minimizing costs and maximizing efficiency when putting environmental protection standards into practice. For one thing, the owner of a facility should pay directly for reducing the pollution it generates: the government should not underwrite these costs (e.g., by providing low-interest loans to finance pollution abatement, by granting special tax advantages, or by paying direct operating subsidies). This principle is advanced despite full knowledge that society as a whole will ultimately have to pay the pollution abatement expenses. The major arguments in favor of this approach are as follows:

1. The polluting firm will have the greatest incentive to adopt the most economical means of meeting the standards if it must absorb the direct costs. This is true even though society will eventually pay those costs.

2. The increased pollution abatement costs that are passed on to consumers may encourage them to use other products involving less pollution,

thereby economizing on the expense involved. For example, a higher copper price may induce a shift to aluminum, whose production requires lower pollution abatement costs.

The second principle on which most environmental economists agree is that environmental regulatory authorities should gear the fees imposed on polluting firms to the amount of noxious substance emitted.[9] For example, smelters might be charged a certain rate per pound for SO_2 released. Firms would then strive to reduce their smelter emissions to the point at which the additional cost of eliminating the last pound would be just equal to the charge per pound. Different firms would reduce SO_2 discharge by amounts depending upon the additional costs involved. Moreover, each firm would be permitted to employ the most cost-efficient type of control, whether by modifying the production process or by capturing the emissions, say, in the form of solid sulfur that could then be marketed. A system of effluent fees could achieve the desired ambient air and water quality standards for copper smelters at a lower social cost than that under present EPA regulations—and without dictating how the standards should be met.[10]

Finally, environmental regulations should aim for an even balance between marginal social benefits and marginal social costs. Although it may not be possible to estimate total benefits or total costs for removing a certain amount of pollution, enough information is usually available to calculate the approximate point at which a further reduction in the level of pollution yields sharply decreasing benefits. Also, in most operations there will be a point beyond which the reduction of an additional unit of pollution requires sharply increased costs. For example, there may be a relatively low-cost method of capturing 90 percent of SO_2 emissions from a smelter, but to capture, say, 95 percent would cost a great deal more. Provided the marginal benefit and cost functions have these characteristics, the net social benefits from pollution abatement are maximized when marginal benefits are just equal to marginal costs. Environmental authorities may not know enough to determine this point and in any event the marginal cost functions will differ substantially from plant to plant. What they could do, though, is require all plants to reduce pollution to the level below which further reductions will yield sharply declining benefits. In addition, they could impose a fee or tax on each ton of pollutants emitted at that level: discharges below the mandated level would reduce the fees the plant had to pay. In this way each facility could reduce its emissions to a level that would equate its marginal cost for reducing pollution to the tax it would be required to pay if it failed to do so. Fees should be structured so that the penalty for releasing an additional unit of pollution approximates the additional social benefits to be gained if the unit were eliminated. This method could prove quite difficult in practice, but it would make possible an approximation of the degree of pollution abatement needed to maximize net social benefits.[11]

Environmental Legislation and Regulations Affecting Mining and Mineral-Processing Industries

The history of U.S. environmental protection legislation and its enforcement is too long and complex even to summarize adequately. The principal concern is whether this legislation will substantially reduce domestic output of basic metals, particularly aluminum, copper, and steel.

The lack of adequate information about pollution abatement costs and about future legislation and enforcement makes objective analysis of this question difficult. Environmental laws and regulations arise from political compromise: they almost never satisfy all interested parties or reflect the social benefit-cost criteria of economists. Legislation is often worded ambiguously, leaving the EPA fairly wide discretionary powers. The actual regulations are powerfully affected by administrative feasibility, court interpretations of legislation, and the values and biases of enforcement officials—themselves often subject to political pressures.

Three categories of environmental controls involve important costs for the nonfuel-minerals-mining and -processing industries—namely, those setting clean air, water, and solid waste disposal standards. Although there is considerable uncertainty regarding the ultimate cost of these regulations, those pertaining to air pollution have so far been the most expensive.[12]

Clean Air

The principal atmospheric standards that affect the mining industry concern SO_2, the major air pollutant from copper, lead, and zinc smelters. The Clean Air Act Amendments of 1970 (Public Law 91-604) does not mention social benefit-cost analysis among criteria for determining ambient air standards. It stipulates only that the EPA set "technologically and economically achievable" goals. ("Economic achievability" for a firm or industry is an ambiguous term, but the EPA has estimated the cost of proposed technological standards for several industries.) The 1970 legislation focuses on control of emissions and ambient concentrations of six pollutants: sulfur dioxide, suspended particulates, carbon monoxide, nitrogen oxide, unburned hydrocarbons, and photochemical oxidants. The head of the EPA is required to establish primary and secondary ambient air quality standards for the nation as a whole. The former establish maximum allowable concentrations in the atmosphere "to protect the public health . . . allowing for adequate margin of safety. . . ." The states are permitted to establish even stricter standards, and they share the administration of them with the EPA.

Requiring the EPA to set *primary* air quality standards to "protect the public health" and *secondary* standards against "any known or anticipated adverse effects" implies that there are threshold levels of pollution below which no damage would be detectable. Critics object that the threshold concept brackets out health damage caused by relatively low levels of pollution over long periods of time, and that the degree of impairment form a continuum, ranging from none when there is no pollution to great damage at high levels.[13] Some environmental economists advocate a system of per-unit emission charges. This would avoid the difficulties of the threshold concept, since all measurable pollution would be taxed; it would also provide an incentive to reduce emissions to a level at which marginal benefits equaled marginal costs.

EPA regulations based on the Clean Air Act Amendments of 1970 have the following objectives: (1) attainment of national ambient air quality standards; (2) use of the best pollution abatement technology in new plants; (3) prevention of significant air quality deterioration in "clean" areas; and (4) progress toward attaining national standards in "dirty" areas. In general, national quality levels must be met by using permanent emission reduction technology, not the intermittent control systems and periodic production curtailments employed previously. Additions to existing capacity as well as all new facilities must use Best Available Control Technology (BACT) for permanent emission reduction. Moreover, the standards themselves derive from what can be achieved by BACT rather than from a comparison of the social costs and benefits involved.

EPA regulations with respect to air pollution are flawed, because the measures they require increase costs more than necessary and cause the unnecessary shutdown of some plants.[14] The most significant criticism applies to new-source performance standards requiring BACT. Both private firms and environmental economists tend to agree that firms should be permitted to use whatever technology they choose as long as the standards are met.

Clean Water

The 1972 Water Pollution Act Amendments set a deadline of 1977 for achieving point-source effluent limitations using the Best Practical Technology (BPT) and of 1983 for limitations using the Best Control Technology (BCT). In addition, the EPA establishes performance standards for new sources. The requirements are effective only if "the administrator finds . . . that such elimination is technically and economically achievable for a category or class of point sources . . ." (Federal Water Pollution Control Act, 1972 Amendment [Public Law 92-500]). Thus attainment of water quality objectives (basically the fishable-swimmable-drinkable standard) depends upon the use of best current practice or best available technology, not upon weighing social benefits against social costs to determine specific quality standards. In general, the same criticisms apply to the technology-based standards of the 1972 act as to those of the Clean Air Act Amendments of 1970.

Solid Waste Disposal

The Resource Conservation and Recovery Act of 1976 (RCRA), as amended in 1984, provides for EPA regulations on disposal of industrial solid waste, with particular reference to hazardous materials (i.e., those that may endanger public health). The mining industry produces about 2.3 billion tons of solid waste each year, falling into three general categories: (*a*) mine waste, including overburden (the surface material removed to get at useful minerals) and subsurface materials extracted along with ore, such as rock, clay, sand, and gravel; (*b*) "tailings" that result from crushing, screening, and concentrating ore; and (*c*) miscellaneous debris created by site preparation and construction of mines and ore-processing plants. Copper extraction produces the largest amount of solid waste, followed by mining for iron, phosphate, uranium, and bituminous coals. Mineral-source solid waste can be hazardous to health and the environment when there is pyrite in tailings and other residue. Pyrite's acid runoff can do considerable environmental damage if it flows into surface drainage systems without treatment. Dangerous residue from copper, lead, and zinc mining results mainly from concentrating and processing crude ore for smelting. Traditionally, waste from these sources is disposed of in tailing ponds that, if untreated for removal of certain metals, constitute an environmental danger. EPA regulations exempt overburden returned to the mine site for surface reclamation, but waste rock and overburden from uranium and phosphate mining are considered hazardous waste subject to special management standards. At this writing, the EPA had not determined what quantities of waste to consider hazardous.

The increased cost of waste disposal under RCRA regulations depends heavily on whether the concentrating and processing residue is classified as nonhazardous or hazardous. Nonhazardous waste disposal is estimated to cost $83 per ton of copper—about 4.5 percent of the June, 1980, price. For lead and zinc the corresponding additional cost is $30 per ton, an increase of 4.3 percent from the June, 1980, average combined price of lead and zinc. If the EPA considered these wastes hazardous, the added cost of disposal would amount to 360 percent of the copper's value, and 153 percent of that for lead and zinc. Should such a reclassification take place, the copper-, lead-, and zinc-mining industries would be put out of business,[15] but there is virtually no possibility that this will occur.

Impact of Environmental Regulations on Major Nonfuel Mineral Industries

A decade and a half after the Environmental Protection Act and the Clean Air Act of 1970, the federal programs are still in an experimental stage and subject to constant change. As serious problems for particular industries have arisen, regulations or laws have been liberalized, but people and groups concerned with environmental protection continue to urge the tightening of standards. This situation is likely to continue for decades because of increasing knowledge of environmental problems (e.g., how legislation to curtail acid rain affects the mineral industries) and because of shifts in the political influence of the parties concerned. Accordingly, the studies on the economic impact of environmental legislation reviewed in appendix 8-1 presuppose that existing legislation is fully enforced; if it is not, their conclusions are likely to be invalid. This is particularly true for studies that project the virtual destruction of an industry or a substantial reduction in its output due to environmental regulations. Such has never been the intent of Congress, which would certainly shy away from disastrous results of this sort.

An important consideration is whether environmental regulations applicable to foreign industries that compete with American producers are comparable in pollution abatement costs. In general, EEC and Japanese standards are similar to those set by the EPA, but for reasons discussed in appendix 8-1, actual costs may be lower abroad. Pollution abatement standards are lower in most developing countries than in the United States, so costs are also correspondingly less.

Copper

Studies by government agencies and private firms have concluded that full compliance with EPA clean air regulations would reduce U.S. copper-smelting capacity and substantially increase American import dependence for copper by 1987. (The principal reason is not that the copper industry is unable to meet the primary quality standards, but that it is unwilling to install permanent controls requiring extensive modernization and new smelting technologies.) These conclusions were based on the assumption that future prices would not pay the added costs of copper production, including interest and amortization for new equipment.

Copper prices have in fact been lower than those projected for the 1980s in these studies made in the late 1970s, and U.S. copper output has declined by nearly one-third from its 1981 level. Full compliance with EPA regulations is required by 1987; by then all the old reverberatory smelters will have been shut down. Some new smelter capacity that meets EPA standards has been installed, and a further reduction of U.S. copper output is not anticipated as a consequence of environmental regulations.

Steel

U.S. environmental legislation required the iron and steel industry to comply with certain air and water quality standards and to install (by December 31, 1982) advanced treatment technology that would have involved substantial new investment. Because of low profits, the industry was judged unable to invest enough to meet the stricter standards and to modernize its plants simultaneously in order to operate more efficiently. In 1981, Congress passed legislation allowing companies to defer the installation of permanent pollution control equipment until the end of 1985 so that they could devote capital to modernization at least equal to the deferred capital expenditures for pollution control.

The impact of environmental regulations on the steel industry's ability to compete with imports is difficult to predict: there are too many uncertainties about required installation of advanced control equipment and the ultimate costs of pollution abatement here and abroad. The government is likely to continue giving the industry special consideration both through import restrictions and through deferral of certain EPA regulations. Therefore, the cost of pollution abatement will probably not be the major factor in determining the U.S. steel industry's viability.

Aluminum

Pollution from aluminum smelting is a less serious public health problem than waste from copper and steel production. In addition to meeting emission standards for existing plants, aluminum companies are held to stricter norms, based on Best Available Control Technology, for new production facilities. Although investment in pollution abatement as a percentage of total investment has been relatively low in the past, it is expected to rise over the next decade and a half as new capacity or additions to existing capacity are constructed. Since there is considerable question regarding how much new capacity will be brought on line, expenditures for BACT may not greatly affect the future competitive position of the aluminum industry. In any event, other items, such as the cost of power, will be much more important in determining future U.S. import dependence for aluminum.

Access to Public Lands for Mineral Exploration and Mining

Perhaps the most important conflict between mining interests and environmentalists centers on the availability of federal lands for exploration and mining. The government owns close to 760 million acres (about half of it in Alaska), or some one-third of all land in the United States. According to the Department of the Interior Task Force on the Availability of Federally-Owned Mineral Lands, about 42 percent of these lands has been completely withdrawn from mineral activity, another 16 percent is severely restricted, and 10 percent is moderately restricted.[16] The major territory at issue is some 80 million acres presently designated as wilderness areas, and about 174 million acres identified by the Bureau of Land Management (BLM) as Wilderness Study Areas (WSA) for possible inclusion in the wilderness area category.[17]

Environmental groups like the Wilderness Society and the Sierra Club seek large additions to the presently designated wilderness areas and advocate strict rules against exploration and mining within any of them or the national parks.[18] The mining industry wants the Wilderness Act of 1964 amended to eliminate its prohibition of exploration and mining in all wilderness areas, including WSAs. It also seeks to limit the amount of land brought into the wilderness system.

How does the availability of federal lands relate to nonfuel minerals availability for the domestic economy? Over the shorter term, say, the next ten to twenty years, the main issue is exploration for and development of minerals (such as chromium, cobalt, and the platinum-group metals) for which the United States depends almost entirely on imports. In most cases, production of these metals would be profitable only if it were directly subsidized or given heavy import protection. The argument for domestic development of these materials is mainly that this would provide greater security against import disruption (see chap. 7).

At issue over a much longer period is exploration to expand reserves of all useful minerals as a contribution to world supplies, and to maintain an "adequate" domestic mineral base and a "strong" domestic mineral industry. At this level, the argument for exploiting wilderness areas to increase supply is both weak and ambiguous, especially for minerals like copper and iron whose reserves are abundant here and abroad. On a global level, the argument is weak because when alternative social values and the relatively low quality and abundance of mineral resources in U.S. wilderness areas are considered, the latter are not very attractive for mining. Moreover, if the depletion of world reserves of certain minerals eventually leads to very high prices, whatever mineral resources now exist in wilderness areas will still be there. The argument is ambiguous because it is not clear why national security depends upon the exploitation of wilderness area minerals.

The Designation of New Wilderness Areas

The Federal Land Policy Management Act of 1976 (FLPMA) directed the BLM to review its holdings in order to select areas suitable for wilderness designation. The law allowed fifteen years, until 1991, to complete the task. The Alaska Land Act of 1980 resolved the wilderness issue for BLM-administered acreage in the forty-ninth state, while the bureau's scattered tracts in the East were largely excluded from consideration by the FLPMA itself. The BLM review basically involves about 174 million acres in the contiguous forty-eight states. The first task was to choose the specific areas with wilderness characteristics: each became a WSA. The WSAs are examined for their wilderness quality, for conflicts with other resources uses such as mining and lumbering, and for management problems that might prevent the BLM from administering the units as wilderness areas. As of 1983, identification of 934 WSAs containing some 2.4 million acres had been accomplished. The Reagan administration accelerated the review process: BLM recommendations for congressional designation as wilderness were to be completed by 1984. Environmental groups objected that the new timetable imposed undue haste.

When the wilderness system was organized in 1964, it took in only 9 million acres, but by 1980 it had grown to over 80 million acres (all in the original forty-eight states). A small amount was added in 1984, and much more is certain to come from WSA acreage not yet definitively classified, but since substantial mineral potential weighs against wilderness area designation, not much of the newly dedicated acreage is likely to be rich in minerals. Nor is existing acreage in wilderness and primitive areas on the whole attractive for mining.[19] These areas have been surveyed by the United States Geologic Survey, and over the years some exploration and even extraction has been carried on by private claimholders. Further exploration would undoubtedly indicate the existence of some additional mineral reserves. A 1982 survey of USGS reports covering sixty wilderness and primitive areas reveals that 60 percent had little or no economic resource potential, 22 percent had moderate potential, and only 18 percent had relatively high potential.[20] The existence of high economic mineral resource potential would usually have prevented parcels of public land from becoming wilderness areas in the first place, since they would probably have been subject to considerable exploration and development under mining claims covering a large part of the area.

How should decisions be made regarding which federal lands should be reserved solely for recreation, wildlife preservation, and other noncommercial uses, and which should be available for timber, mineral, and other economic development? For the most part, the current debate involves a clash over absolute values—national income and security versus national environmental amenities. A rational approach to the problem lies in quantifying the value of contributions to national income and security on the one hand, and to environmental amenities on the other, followed by the comparison of these values through social benefit-cost accounting.

The economic value of services of land used for extraction—including forestry, agriculture, and mining—can be estimated from the discounted streams of the product's net values (i.e., their market values less extraction costs). Determining the contribution to national security gained by devoting land to exploration and mining presents more complex valuation problems. One approach would be to compare the cost of increasing the domestic supply of a critical mineral in this manner with the cost of providing the same benefit by an alternative means, such as stockpiling. It is important, though, to realize that it is impossible to quantify vague benefits such as a "strong domestic minerals industry."

If using an area for extractive purposes would sacrifice important environmental amenities, the social cost of such use includes the value of "services" lost through commercial exploitation. But these lost services (recreational benefits and preservation of wildlife, both animal and vegetable) normally have no commercial value; their value derives mainly from the public good they serve. What is required, therefore, is a method for calculating the value of the services derived from leaving an area in its natural state.

The Valuation of Environmental Benefits

Over the past fifteen years, environmental economists have formulated conceptual approaches to social benefit evaluation of wilderness and scenic areas. It is, of course, harder to quantify the recreational value of forests, canyons, wild rivers, and the like than to evaluate benefits from clean air and water and the elimination of hazardous waste. (However crude and controversial it may seem, monetary costs can reasonably be assigned to impairment of health and increased mortality.) One method of fixing a price for the more ineffable benefits under consideration here involves calculating the net travel costs incurred by visitors to a wilderness area. This approach would provide a minimum estimate of the dollar value of recreational experience. Another method is to interview various classes of visitors, including hunters, fishermen, car campers, and backpackers to ascertain the maximum fee they would pay to visit the area.[21] The value of future recreational benefits could be estimated by projecting the number of visitors each year in the future and calculating the annual recreational values. The present value of the estimated recreational benefits could then be determined by discounting the stream of annual benefits by the social rate of discount. Choice of the social rate of discount raises a social welfare problem; its importance is indicated by the fact that the lower the rate of discount is, the higher the present value of recreational benefits.[22] The discounting process raises the issue of whether evaluation should apply solely to benefits for the present generation of consumers or should also take future generations into account. If the benefits of future generations and of the present generation are regarded as equally important, no discounting of future benefits should be employed. However, few people living today would be willing to equate the value of their wilderness experience with that of someone living in the year 2050.

The evaluation of social benefits from wilderness areas should ideally include the "optional value" that some individuals derive from knowing that unspoiled wilderness exists, even if they never visit them. Or they may strongly desire to preserve these options for others, including future generations. No feasible method has been developed to determine optional values. Still other social values from wilderness preservation cannot be measured by direct recreational use. Examples include the protection of ecological systems, the preservation of biological species, and the availability of botanical specimens for scientific knowledge.

The concept of *irreversibility* is frequently employed in estimating environmental costs of mineral development.[23] Building roads and mining communities in the center of a wilderness area is likely to damage it beyond repair even if pits are filled, structures removed, and mining sites reforested later on: the area's remoteness and ecology will never be the same. This may also apply to intensive exploration, even though no economical mineral deposits are discovered. Since exploration leads to the exploitation of commercial deposits in only one of twenty or more ventures, the value of what will be destroyed forever must be weighed against the probability-adjusted social value of minerals to be extracted. Over the longer run, technology may develop substitutes for scarce minerals found in a wilderness area, but it cannot replace lost environmental amenities. Moreover, as the population grows, those amenities are likely to grow much faster in value than the scarce minerals that may be replaced by more abundant materials. A corollary of the irreversibility principle is that preserving the environment and leaving minerals in the ground means they will be available later should the country (or the world) need them desperately.

Although the idea that social benefit-cost accounting is pertinent to environmental issues has won a substantial following, practical application lags behind. In part this is true because of a lack of information, in part because both environmentalists and their opponents are skeptical about the method. Economists concerned with environmental issues believe that environmentalists pass up a valuable opportunity when they decline to present their case in monetary terms. Certain development activities in wilderness areas can often be shown to make no economic sense even without figuring in environmental costs. (This has clearly been the case with a number of water projects, and I suspect it may also be true for most, if not all, commercial development in wilderness areas.) On top of this, a demonstration that monetary costs of environmental destruction greatly exceed possible economic benefits might be more convincing to the public and its congressional representatives than emotional appeals alone.

A case in point is a study that estimated the present value of recreational services in the White Cloud Peaks area (a WSA) in Idaho's Challis National Forest.[24] It compared this figure with the estimated present value of ASARCO's claim to a low-grade molybdenum deposit that, it was assumed, could not be profitably exploited for thirty years.[25] The study calculated the present value of recreational benefits based on projected growth in recreational demand. The present value of the mining venture was estimated by projecting ASARCO's net annual returns.[26] Present values for the alternative uses have similar magnitudes.[27] The study holds that even if the mine's net present value were greater than that of the alternative, incompatible recreational services, the irreversibility principle, the uncertainty of supply and demand for the minerals, and other unmeasurable factors would tip the decision in favor of wilderness preservation. (The White Clouds Peak area, incidentally, is one of those under BLM consideration for wilderness area status.)

Conclusions

The primary metals industries, especially iron and steel and nonferrous metals, are among the largest polluters of air and water in the United States. The mining industry is in conflict with environmental groups over pollution abatement, solid waste disposal, and the proper use of public lands. How these issues are resolved is of great importance for national welfare. A rational solution involves comparing the social benefits of environmental preservation with its social costs. Unfortunately, neither mining spokesmen nor environmentalists have favored this approach. Nor has it been sanctioned in congressional legislation. Nevertheless, the principle of comparing social values and costs is implicit in many public policy decisions on the environment. What is lacking is a more rigorous, explicit application of social benefit-cost accounting.

Pollution abatement costs should be borne initially by the polluting firm, even though society will ultimately pay. Otherwise companies would have no economic interest in using the most cost-efficient method of reducing pollution. Since a program of complete pollution avoidance involves unacceptable social costs, a trade-off is necessary between the additional social benefits generated by further reductions in air, water, and land pollution and the social costs they would entail.

The principal criticism of EPA regulations to alleviate air and water pollution has not been with the quality standards expected of air or water exposed to nonferrous smelters, blast furnaces, and aluminum plants as such, but of the technical requirements for achieving these standards. In general, new plants and additions to old plants must employ Best Available Control Technology rather than the less expensive technology previously used. Moreover, the emission standards themselves are based on what is technically possible rather than on the environmental impact of pollution itself. A number of economists urge that it would be more economical to charge firms according to the quantity of pollutants emitted, since this would give them an incentive to spend more on reducing pollution—at least until additional costs equaled the charges for additional emissions.

The capital costs of new pollution abatement equipment required by EPA regulations constitute a substantial percentage of total capital expenditures for steel and nonferrous metals industries. These costs, together with direct operating costs for pollution abatement, reduce the competitiveness of U.S. industries if they greatly exceed analogous expenses abroad. This does not mean that polluting industries should be exempt from these costs, especially in view of the social benefits from cleaner air and water and the elimination of hazardous solid waste. It does mean that no higher cost or reduction in productivity is justified than what is necessary to meet the regulatory standards.

Debate over the availability of wilderness areas (and of areas that might qualify as such) for exploration and mining has been carried on without relevant facts or objective analysis. Conducted largely in terms of private interests and alleged national security benefits on the one hand, and broad appeals for preservation of increasingly scarce wilderness environments on the other, a rational resolution of the issue has been avoided. Application of social benefit-cost accounting is difficult here because of the lack of appropriate methodologies and data adequate to measure environmental services. Neither of these difficulties is insurmountable, but progress will require wider acceptance of the social benefit-cost approach by environmental groups as well as the government and the general public.

Whatever the outcome of social benefit-cost studies in cases involving alternative uses of particular tracts of public lands, the argument that national welfare requires increased domestic production of nonfuel minerals to assure their domestic availability is unconvincing. As the previous three chapters have shown, alternative measures are adequate to protect the economy against import disruption of needed minerals. Even if there is something to the claim that a strong domestic minerals industry is important to national security and economic welfare, the industry currently has no shortage of areas to explore and exploit. If world supplies were eventually depleted, prices soared, and substitutes were not feasible, wilderness area minerals would still be there, but destruction of wilderness amenities is irreversible.

Appendix 8-1

The Impact of EPA Regulations on the Copper, Steel, and Aluminum Industries

The Copper Industry

EPA regulations mandate that the national ambient air quality standards for SO_2 be met by the use of permanent emission reduction technology, not by tall smokestacks and production cutbacks during adverse meteorological conditions. All new capacity, including additions to existing plants, must use BACT. The Clean Air Act Amendments of 1977 permitted smelters unable to comply with the regulations requiring permanent emissions technology to operate under a five-year variance, *renewable once,* provided each smelter installs as much pollution control equipment as it can afford.

Technically, the best current method to capture SO_2 emissions from smelters is in the form of sulfuric acid, which has a variety of industrial uses. Firms in Japan and Western Europe can market sulfuric acid obtained in this way at a price that covers the cost of capture. But in the United States, which has natural sources of sulfur as well as additional supplies from natural gas and crude oil desulfurization, sulfuric acid produced by smelters is not in great demand. Also, because shipping costs to sulfuric acid markets are high, domestic producers have been unable to dispose of smelter-produced sulfuric acid profitably.[28]

In 1976 the EPA contracted with Arthur D. Little, Inc. (ADL), to study the economic impact of EPA regulations on the copper industry, with particular reference to SO_2 smelter emission. The ADL study, submitted in January, 1978, found that emission limitations will cost existing smelters a great deal in the short run and will rule out the incremental smelter capacity expansion traditional in the industry.[29] ADL projects that higher domestic prices will cause imports to grow substantially and U.S. copper consumption to decline. (Extrapolation of domestic prices indicates a rise of some 29 to 39 percent by 1987 because of the increased costs.) The additional expenditures needed to maintain existing capacity between 1983 and 1988 may cause selected plant shutdowns. The study anticipates that by 1987 domestic copper production will fall by nearly 25 percent below a baseline expansion projection, even if the industry were permitted to meet the national air quality requirements through minimum control devices with no constraints on the technology employed. Finally, since expansion at the mining, milling, and refining stages of production is closely linked to the growth or contraction of smelting capacity, the other three stages will be limited in growth by the decline of smelting.

How accurate have the ADL study projections proved to be by the end of 1985? First, the study's baseline projection of the U.S. copper industry's normal growth was substantially overstated, because copper consumption did not grow at the projected level during the 1980s. Copper consumption in the United States declined from a peak of 2,433,000 mt in 1979 to 1,760,000 mt in 1982 and was about 2,100,000 mt in 1984 and 1985.[30] Between 1979 and 1985, U.S. copper mine production fell by about 27 percent, and domestic and world prices declined by some 30 percent. Refined copper imports were approximately the same in 1984 as in 1980. There has been a decline in U.S. copper smelting capacity, because several old reverberatory smelters that did not meet EPA standards were closed; the remaining reverberatory smelters are scheduled to be shut down in 1987. Some capacity would have been shut down even in the absence of EPA requirements, since both the smelters and the mines associated with them were uneconomical at prevailing copper prices.

The only copper smelter built in the first half of the 1980s that meets EPA standards for new capacity is that of the Chino mine in New Mexico, owned by Kennecott Minerals Corporation (a subsidiary of Standard Oil of Ohio).[31] This facility is a flash smelter, which replaced the old reverberatory furnace.[32] To meet EPA standards, the reverberatory smelters will have to be replaced with flash smelters or hydrometallurgical processes that do not emit SO_2.[33] Some new leach-electrowinning facilities (a hydrometallurgical process) were installed during the first half of the 1980s. However, it is projected that by the late 1980s the United States will not have sufficient smelting capacity for all its projected mine output and that about 175,000 mt of copper in concentrates will be shipped overseas.[34] On the other hand, if copper prices rise, new facilities for producing copper from concentrates are likely to be built. Moreover, a rise in copper prices is likely to result in reopening of closed mines with a total capacity of about 600,000 mt.[35] This could raise production from about 1.1 million mt (1985) to about 1.7 million mt.

EPA regulations have been responsible for only part of the drop in American mine and smelter production from its peak of 1.5 million mt in 1981 to about 1.1 million mt in 1985. The principal reason for the decline has been the fall in copper prices, which has rendered a substantial portion of U.S. capacity unprofitable.

The Iron and Steel Industry

The iron and steel industry produces several categories of pollutants at each stage, from the preparation of iron ore suitable for charging into blast furnaces to the production of finished steel products. The pollutants include airborne particulates and gases containing sulfur and nitrogen oxide, waste water contaminated by chemicals and suspended solids, and heavier materials such as sludges containing hazardous substances. Various forms of control equipment are employed, depending upon the technology used in production (e.g., open hearth furnaces versus electric arc furnaces).[36]

The Clean Air Act Amendments of 1977 established a deadline of December 31, 1982, for meeting EPA pollution control requirements for the steel industry. The Steel Industry Compliance Extension Act of 1981 permits the EPA to extend the date on a case-by-case basis for up to three years, that is to December 31, 1985. (According to an EPA official, most steel firms were in compliance by late 1986.) The extension's purpose is to enable steel companies to defer pollution control expenditures in order to invest more capital in production itself. (Within two years of the 1981 law's enactment, an amount equal to the deferred pollution control expenditures must be used for additional investment in operations.) The law further states that the stay must not result in air quality degradation and that companies must make continued progress toward compliance during the extension period. However, as of mid-1984 the steel industry had spent only $49 million on modernization, in contrast to the $500 million to $700 million it saved by deferring investment for pollution abatement.[37]

According to a congressional committee report, the steel industry was given special treatment because of the difficulties it faced from foreign competition and because this industry must devote a greater percentage of its capital expenditures to pollution control than any other major industry.[38] For example, 19.0 percent of iron and steel companies' total capital expenditures went for pollution control equipment during 1980. This contrasts with only 3.9 percent for all industries, and 9.4 percent for those producing nonferrous metals.[39] The report also pointed to the relatively low rate of return on equity in the iron and steel industry during the 1979–80 period—8.5 percent as contrasted with 15.4 percent for all industries.

Resources for the Future, Inc. (RFF), published a study that projected investment in pollution abatement as a percentage of total investment by the steel industry for 1981–85 and for 1986–90. Assuming relaxed EPA standards, these percentages were 5.9 and 13.0 for the 1981–85 and 1986–90 periods respectively. Stricter standards based on advanced treatment technology could mean a rise to 16.8 percent for water (1983) and 14.7 percent for air (1980) and to 27.4 percent of total investment for the 1991–95 period.[40] Although investment plans based on these projections changed radically as a consequence of the 1983–84 steel industry restructuring, the RFF figures probably give a good indication of the ratio of pollution abatement to total investment that EPA standards dictate in the industry. However, not much of the required antipollution investment was made in 1981–85.

There is evidence that pollution abatement costs less abroad than in the United States. A 1977 Organization for Economic Cooperation and Development (OECD) study compared U.S. emission control expenses with those for five European countries (Belgium, Finland, France, West Germany, and Sweden).[41] U.S. costs (as of 1975) were $19.85 per ton of steel as compared with $6.60 for Germany, $5.22 for Sweden, $2.09 for France, and levels for Belgium and Finland similar to that of France. This report also compared emission control expenses for the six countries assuming the steel industries in all six had maintained the same standards. Under that assumption, the cost impact per ton of steel was only slightly higher for the United States than for the European countries.

Meeting EPA standards has imposed a substantial cost burden on the U.S. steel industry, probably a greater burden than that borne by its foreign competitors. It is not possible to project the cost of pollution abatement with any confidence, because the status of regulations for installation of permanent control strategies remains in flux. One estimate is that total capital and operating costs for pollution abatement will eventually rise to about 5.5 percent of product prices in the absence of control.[42] However, there could be considerable savings if the industry were permitted to employ lower-cost strategies for the same reduction of pollution emissions.[43] In any event, pollution abatement will not be the major influence on the future viability of the U.S. steel industry. (The most important factors in the industry's declining competitiveness were given in chap. 5.)

The Aluminum Industry

The most important pollutants generated in making aluminum are the fluorides produced in both gaseous and solid form during the reduction (smelting) process. Total fluorides generated differ with the technology employed, but amounts range from twenty-seven to sixty-five pounds per ton of aluminum, with the largest component in gaseous form. Gaseous emissions include a number of other pollutants besides fluorides.

The EPA has determined that fluoride emissions from primary aluminum plants may cause or contribute to public endangerment,[44] although adverse effects on public health have not been demonstrated. This means that fluoride emissions will be considered a welfare-related pollutant and that states will have greater flexibility in establishing fluoride control guidelines than they would if public health were clearly affected.

The EPA has set maximum amounts of fluoride in gases that may be discharged into the atmosphere for each major type of aluminum facility. The standards are couched in terms of pounds of fluroide per ton of aluminum produced. Capital costs for equipment required by *existing* facilities to meet the standard vary considerably by type of plant. For one representative system, the capital cost is estimated at $386 per ton of aluminum.[45] Estimates of additional annual operating expenditures are available for only a few types of aluminum plants; they range from $0.81 per ton to $9.57 per ton of aluminum.[46]

Performance standards for *new* aluminum reduction facilities are uniform nationwide, based on "best control technology" that has been "adequately demonstrated"—meaning that the technology and the test data are available. Additions to or modifications of existing plants must also meet new source standards. However, states may set stricter standards if they so desire. During the period 1976–80 investment in pollution abatement as a portion of total aluminum industry investment ranged from 9.4 to 10.1 percent. The proportion is expected to be lower—4.4 to 8.3 percent—for 1981–85 but is projected to rise steadily to a range of 20.9 to 28.5 percent in the 1996–2000 period as new capacity or additions to existing capacity are constructed.[47]

In most other developed countries, regulations on fluoride emissions by the aluminum industry are either more liberal than EPA standards or nonexistent. However, a Department of Commerce report concludes that "while abatement costs contribute to production cost increases, it appears unlikely that the differential cost between countries will dislocate the present international trade patterns in primary aluminum."[48]

9

Outlook and Policy Conclusions

Mineral Resources and the Limits to World Growth

Public policies on major nonfuel mineral issues should be formulated according to the prospective relationship between mineral availability and world economic growth. If the outlook is for an imminent resource scarcity that would jeopardize national and world economic progress, drastic programs to curtail the use of specific minerals would be in order. However, this study finds that nonfuel mineral supplies should be ample to meet world demand for the remainder of this century and well into the next.

Declining rates of growth in the demand for metals, together with an increase in reserves and identified mineral resources, virtually guarantee that enough nonfuel minerals will be available to meet world requirements—without sharp rises in real prices—until 2000. There may be inadequate capital and technology to fully develop these resources in some LDCs, but this is likely to be offset by greater production in developed countries. The generations beyond 2000 must look to the discovery and proving of additional reserves from the large volume of speculative and hypothetical resources, plus eventual exploitation of identified resources that are currently not economical. Continued technological advances in exploration, metallurgy, substitution, and conservation are also important, both for expanding the mineral reserve base and for reducing the ratio of nonfuel mineral consumption to GNP. A review of known and probable global resources plus a projection of advances in technology provides reasonable assurance that nonfuel mineral scarcity need not interrupt world economic progress until 2030 at the earliest.

The mid-twenty-first century is well beyond the range of projections based on existing knowledge of mineral resources or technology. Nor is it possible to project with any confidence what world production of goods requiring nonfuel minerals will be in 2050, since these are by no means the only potential constraint on economic growth. (Energy and food production also qualify, for example.) Starting with the belief that human ingenuity will continue to solve technical problems in the next century as well as it has in this one, it does not greatly stretch the imagination to foresee almost unlimited substitution of nonfuel minerals (e.g., aluminum and silicon) that constitute a high percentage of the earth's crust for the less abundant ones. The recommendation that developed countries limit their consumption in order to conserve nonfuel minerals for future generations will go unheeded. And so it should, for no-growth policies are likely to retard technological progress—the world's best hope for avoiding material constraints on growth.

The above conclusions are especially relevant to the debate between those who think a scarcity of natural resources inevitably limits world economic progress, and those who believe mankind can overcome the constraints of finite natural resources and can continue to grow in numbers and material well-being for the indefinite future. Division on this issue extends back at least to Thomas Malthus, but the modern debate began with the 1972 publication of the Club of Rome's report *The Limits to Growth,* followed by its most recent counterpart, *Global 2000 Report to the President,* in 1980.[1] The latest of many challenges to these pessimistic forecasts is *The Resourceful Earth,* a compilation of essays by twenty-five leading specialists on natural resources, agriculture, and demography.[2] Although the present study takes no position on other aspects of this global debate, it emphasizes that nonfuel mineral resources need not limit world growth in the foreseeable future.

This study's optimistic conclusions derive much more from the evidence presented than from my own economic philosophy. History and current trends supply the most persuasive evidence about the declining trend of real prices for nonfuel minerals over the past century; the cumulative growth in technology for the discovery, production, substitution, and conservation of minerals; and the record of domestic and world competition in adjusting to economic and environmental changes. Projection into the future is in part a matter of faith, but a faith well grounded in evidence from the past and the present. The findings presented here do not promise mineral abundance for future generations—they only suggest what is possible.

The Need for Mineral Exploration and Development in LDCs

The existence of ample mineral resources in the earth and on the ocean floor does not assure their adequate supply. These resources must be intensively explored and exploited. As surface deposits are exhausted, larger amounts of capital and new technologies are required to discover, extract, and process enough minerals to meet the world's growing requirements. Early in the present century, multinational mining companies revolutionized the minerals industry by exploring throughout the world and by mobilizing large amounts of capital and new technologies to utilize lower ore grades. In doing so, they greatly expanded the availability of minerals needed for the modern industrial economy's enormous growth. Much of the multinationals' investment took place in LDCs, which still possess a very large potential for increased production from their abundant resources.

During the 1970s, many multinational mining properties in LDCs were expropriated, new foreign investment in mining there declined sharply, and state enterprises came to own and control a large share of these countries' mining industries. The state mining enterprises have been heavily underwritten by government-guaranteed loans from international capital markets, but the outlook for new financing from this source over the next decade is not favorable. A major question, therefore, is whether developing countries will have the capital, and in some cases the technology and managerial capacity, to explore and develop their mineral resources in line with the growth in world demand. Adequate expansion in the LDCs would require very large capital investments—on the order of at least $5 billion per year to the end of this century, and rising thereafter. A substantial portion of the LDCs' need for capital and know-how could be provided by multinational mining companies, but they are unlikely to do so unless the investment climate improves.

A continued low level of exploration and development in Third World countries will not necessarily reduce the growth of world output for most nonfuel minerals in the present century, but it may mean that a larger share of it will come from developed countries such as Australia, Canada, and South Africa.

Import Restrictions and the U.S. Competitive Position in Nonfuel Minerals

America's competitive position in major metals—aluminum, copper, iron and steel, lead, and zinc—has been declining over the past thirty years. This decline is likely to continue as the higher-grade mineral resources are exhausted and structural changes take place in the world economy that, together, reduce this country's comparative cost advantage. The structural changes result from relatively high U.S. labor costs, LDC industrialization, and (in the case of energy-intensive metals such as aluminum) high U.S. energy costs compared to those in certain other countries.

These long-term trends must be distinguished from shorter-term developments that have impaired the U.S. competitive position in copper and steel over the past several years. The secular decline in world demand for both nonferrous metals and steel products has created a temporary global overcapacity for these metals. But as world demand begins to grow again and recovery from the worldwide recession of the early 1980s proceeds, the output and competitive strength of U.S. metals industries will increase. Plant modernization and reorganization of the metals industries currently in progress will accelerate this trend. For these reasons, I have rejected the forecast that without protection against imports the U.S. steel and copper industries will be reduced to half or less of their present capacities by the early 1990s.

Import protection to promote a high degree of national self-sufficiency for any industrial branch is self-defeating. By keeping domestic prices higher than world prices, protection inevitably damages the competitive position of all industries that use the protected products. For instance, import restrictions on semifinished steel sap the competitive strength of those who make a variety of finished steel products and of the domestic users of those products, such as the automobile and machine tool industries. Quota restrictions or tariffs that keep domestic copper prices above world levels reduce the ability of copper fabricators and their customers to compete abroad and with foreign imports in the domestic market. Finally, import restrictions reduce the international competitive strength of the industries they are supposed to help, for protection delays needed modernization and restructuring. Economists have been teaching these lessons for more than two centuries, but while there is broad acceptance of the doctrine of free trade, it is often rejected in its application to particular industries.

International Cooperation on Nonfuel Minerals Issues

Several initiatives have addressed international cooperation to deal with nonfuel mineral production. The most important have attempted to reach agreement on stabilizing mineral prices, the control of seabed mining, and public assistance for LDC mineral development. Each of these undertakings is highly controversial, and respected scholars have taken opposing positions on them. I therefore present my own opinions in this area without the same confidence that I feel about the other conclusions in this study.

Fluctuations in nonfuel mineral prices contribute to the instability of investment in new capacity and in the search for new reserves. Unstable prices surely increase the real costs of production. Short-term price fluctuations serve no fundamental economic purpose; moreover, large swings create risks and disutilities for both producers and consumers. Such fluctuations are particularly disadvantageous to developing countries whose foreign exchange income depends heavily on exports of one or two nonfuel minerals. Unfortunately, efforts to reduce the magnitude of short-term price variations for primary commodities have not proved successful. The reasons are in part technical, but more important, price targets and mechanisms for stabilization are determined by representatives of producing and consuming countries with different economic objectives and ideologies. A prime example is UNCTAD's effort to establish an integrated program of commodity agreements. UNCTAD's scheme has been ensnarled in a North-South dispute over Third World objectives that form a part of the NIEO. Therefore the outlook for successful price stabilization in nonfuel minerals is not favorable.

The economic feasibility of seabed mining has yet to be established. It does seem likely that as land-based reserves are depleted, the recovery of seabed nodules will eventually become both economical and highly important for cobalt, copper, manganese, and nickel. But this source is unlikely to become essential for adequate supply before the end of the century.

Recovery of seabed minerals is fraught with both technical and political problems, but the latter may prove the more intractable. The LOS Convention reflects the objectives of the NIEO, and many of its provisions are unacceptable both to the U.S. government and to mining consortia that have undertaken exploration and research for extracting minerals from the ocean floor. A fundamental issue for Washington is whether to ratify the LOS Convention or establish a legal framework whereby American companies can undertake seabed mining outside the convention. Both alternatives present disadvantages and uncertainties as to whether progress toward the establishment of a seabed-mining industry will occur in timely fashion. My tentative conclusion is that the government should not ratify the LOS Convention in the near future; I base this judgment mainly on my conviction that provisions of the present LOS Treaty would impede, if not deter, participation by U.S. firms in the exploration and development of seabed minerals. These provisions may also constitute a barrier to significant development of these resources, in which case the treaty is likely to be amended. In any event, American consumers stand to benefit from increased mineral supplies no matter which national or international entities provide them.

The decline in foreign private investment in the LDCs and the uncertain outlook for private external debt financing of government mining enterprises in those nations have led to the recommendation that international development agencies, such as the World Bank Group, provide the capital and technical assistance needed for adequate mineral resource development. These agencies have supplied little such assistance relative to the large financial requirements needed (see chap. 3). The role of public development agencies in promoting Third World mineral development is a subject of considerable controversy. Both private mining companies and the U.S. government strongly oppose financial assistance to government mining enterprises. They argue that such financing constitutes a subsidy to competitors of American mining firms. This extremely difficult issue has implications not only for international development agency financing of LDC mining enterprises, but for assistance to any LDC export industry that competes with American products.

Clearly, Washington should not try to keep agencies like the World Bank from making loans to promote the exports of developing countries. On the other hand, these agencies were designed to supplement the flow of private international capital, not to finance projects for which international capital is available. Private international capital will undoubtedly be accessible to those developing countries that provide a favorable investment climate. Most Third World countries argue, however, that they have the right to determine how their mining industries should be organized and that they should not be required to accept foreign private investment in order to obtain at least part of the external capital they need. In assessing this issue it is important to recall that the resources of the international development agencies are small in relation to the financial requirements of their client countries. The agencies can supply only a fraction of the many billions of dollars required each year for warranted development of LDC mineral resources. Therefore, the financial resources of international development agencies (whether equity or loans) may find optimum allocation when used as a catalyst to expand the flow of foreign private capital.

Multinational cooperation among the major non-Communist industrial countries to reduce supply disruption vulnerability is more promising, since little if any conflict of national objectives or ideologies is involved. Such cooperation might take two forms. One of them might be an agreement not to impose export restrictions on any nonfuel mineral. This would assure that a contraction of supply from any source would have an equal impact on all countries party to the agreement. Thus, if world supplies of copper or iron ore from South America were curtailed, the United States and Canada (together self-sufficient in these minerals) would have no advantage over Europe and Japan, which lack self-sufficiency for either commodity.

A second form of cooperation could well be the establishment of economic stockpiles by the non-Communist industrial countries, with all inventories coordinated and operated according to the same guidelines. Such stockpiles could equitably distribute the burden of protection against supply disruptions of critical minerals like cobalt, chromium, and manganese. The major industrial countries urgently need to reach agreement on how to deal with the chaotic state of international steel trade. Such an agreement should not further limit international competition by allocating parts of the U.S. market for steel products among foreign producers. Rather, it should define fair trade practices for both exports and imports, and authorize the GATT to establish machinery for adjudicating disputes between parties to the agreement. A sectoral trade agreement in steel products negotiated by the major non-Communist industrial countries would not conflict with their national policies, since all agree with the principles of free trade. Such an agreement could benefit every country, by contrast to the current pattern of penalty restrictions and other retaliatory measures. Once negotiated, the rights and obligations provided by the agreement could be shared by Third World exporters willing to abide by it.

Import Dependence and Vulnerability to Import Disruption

For many basic products, such as nonfuel minerals, protectionists respond to the free trade argument that unrestricted imports increase import dependence.

Import dependence is frequently associated with vulnerability to import disruption, but in most cases this association is fallacious. Except during a general war that interferes with many supply sources and transportation routes, products like copper are available from many sources. This provides greater security than supply from the United States alone, where strikes of many months are common in the copper industry. Vulnerability to import disruption in peacetime concerns only a handful of minerals, such as chromium, cobalt, manganese, and the platinum-group metals, all of whose supply sources are heavily concentrated in areas of political instability.

The issue of nonfuel mineral availability during an all-out war is a complex one, in part because of uncertainty over the kind of emergency for which this country should be prepared in the nuclear age. For some types of emergency, which may be more probable than the type faced during World War II, stocks of finished products at strategic locations may be more advantageous than supplies of raw materials. As for nonfuel minerals in which the United States is substantially self-sufficient (e.g., copper, lead, and zinc), the bulk of requirements could be met during a long conventional war by supplementing domestic production with more of what this country normally imports from Canada and Mexico. For nonfuel minerals (like chromium and cobalt) that are neither produced domestically nor available from contiguous sources, the national defense stockpile constitutes the most economical means of assuring supplies should all foreign sources be shut off.

Manufactured materials like steel products cannot be readily stockpiled because of their high value, large variety, and continually changing composition. It is important for the United States to have a large, modern steel industry capable of meeting the ever-changing requirements of both civilian and defense industries at world prices. There is no evidence that it would be impossible to maintain this kind of steel industry without import restrictions or direct subsidies. Special treatment would be counterproductive in any case, for it is also important to have strong machine tool, transportation, and other industries that use steel products. The strength of these industries depends in part on their ability to obtain steel products at competitive world prices. The essential point is that, in an interdependent world economy, import protection for products used by other industries serves to weaken the entire industrial base. Comprehensive industrial self-sufficiency is not possible except at an enormous cost to the national product and to industrial vitality.

The Need for an Economic Stockpile

The most serious threat to the availability of nonfuel minerals for the economies of industrial nations is disruption of supplies of a handful of minerals, the production of which is concentrated in a few countries highly vulnerable to limited wars or civil disturbances. The national defense stockpile does not protect civilian production from such contingencies. The most effective methods of dealing with this threat would be the creation of a government economic stockpile, or a program to increase private inventories to levels regarded as socially optimal in relation to the social costs of maintaining them. Although the latter approach has certain advantages over the former, it is difficult to say which is the more feasible without a certain amount of experimentation.

The emergence of the OPEC cartel and the Arab oil embargo of 1973–74 gave rise to the fear that foreign cartels, embargoes, limited wars, or civil disturbances may disrupt world supplies of critical nonfuel minerals. This concern is reflected in hundreds of books and monographs and in congressional studies and hearings. Most of this literature is of an alarmist nature and lacks rigorous analysis either of the nature of vulnerability or of methods to deal with a supply disruption. The few studies that have analyzed the problem in a sophisticated manner have either gone unpublished or are so technical that few of those involved in policy-making read them. Despite the widespread concern over vulnerability, no specific proposals for an economic stockpile or the subsidization of private inventories have emanated from the president or the Congress. (Congressional legislation has been proposed for subsidizing domestic production of such nonfuel minerals as cobalt, but fortunately the legislation has lacked administration support.) As with other public expenditures, the creation of an economic stockpile or a program to subsidize private inventories should be justified by social benefit-cost accounting. Since the issue concerns avoiding or moderating the social costs of a supply disruption that is uncertain both in terms of likelihood and magnitude of impact, determining which minerals to stockpile at what levels is a technically complex problem. Nevertheless, rational procedures such as those outlined in chapter 7 have been developed and could be improved upon with experimentation. Even a modest program for a few of the minerals most vulnerable, like platinum, chromium, cobalt, and manganese, and costing no more than $2 or $3 billion, might avoid social costs running to tens of billions of dollars.

A government economic stockpile program or private inventory enhancement would be much cheaper and more effective than subsidizing domestic production. In addition, security could be enhanced by government support of R&D to develop substitutes for, or conservation of, critical minerals subject to import disruption. Government outlays for this purpose can be justified because the potential social benefits far exceed the private gain for individual firms from such outlays.

Mineral Availability and the Environment

Increasing the availability of minerals and preservation of the environment are not incompatible. Nor is it true that economic growth must lead to higher levels of water and air pollution. In many areas of the country, air and water quality have improved over the past decade even as production of minerals and other commodities expanded. Moreover, the mineral industries have reduced pollution without increasing the real prices of minerals; given advances in pollution abatement technology this trend can continue.

Although both pollution and its curtailment involve social costs, the costs of avoiding environmental contamination should be carried in the first instance by polluting industries and should not be offset by public subsidies. The best way to achieve a reduction of industrial pollution is for the government to establish maximum emission standards and impose per unit financial penalties on all discharges. This approach would not only provide a financial incentive to reduce pollution below the mandatory level, but would also stimulate the search for cheaper and more effective means of pollution abatement.

The practice of evaluating public lands solely in terms of the marketable commodities and services they produce should be changed by an explicit recognition of the value their environmental amenities provide. The recreational and ecological utilities yielded by wilderness areas are high and growing, and these social values should be taken into account in deciding on alternative uses of public lands. Such decisions should, wherever possible, be based on social benefit-cost accounting. Given the increasing recreational use of unspoiled public lands, their designation and maintenance as wilderness areas should not be determined simply by whether they contain, or are believed to contain, mineral reserves or harvestable timber. In many cases quantitative studies will show that the social benefits yielded by environmental amenities exceed the probability-adjusted market value of alternative commodity production. Moreover, commercial development destroys the environmental utilities of such areas forever, whereas the potential value of their commercial products remains if their pristine location is undisturbed. Contrary to statements by critics of environmental protection, this approach does not sacrifice economic progress to conservation, since meeting the demand for environmental amenities constitutes an important and growing element in the social product.

Notes

Chapter 1

1. For an excellent statement of this issue, see Hans H. Landsberg (senior fellow, Resources for the Future), "Key Elements Common to Critical Issues on Engineering Materials and Minerals," in Congressional Research Service, *Seventh Biennial Conference on National Materials Policy* (Washington, D.C.: U.S. Government Printing Office, March 1983) (report prepared for U.S. Congress, House, Committee on Science and Technology, 98th Cong., 1st sess., 1983, pp. 11–28).

2. Ibid., p. 26. This point was made in an address to the Seventh Biennial Conference on National Materials Policy: "One of the cherished goals of the materials community is to have a 'comprehensive national materials policy.' I suggest to you that this is an illusive goal at best and a waste of time at worst. It tends to produce little more than the kinds of pious and anemic generalities in which past Congressional legislation abounds. Nor could it be otherwise. The moment you advance from generalities, such as the call for a 'sound and stable minerals industry' (whatever that might mean), you are faced with the fact that in policy terms, iron ore, bauxite, titanium, and copper, just to name a few, have too little in common to be blanketed by a unified national policy that goes beyond exhortations and admonitions."

3. The argument that the national product would be lower if some human and capital resources were no longer employed in the mining industry is invalid in the context of free mobility of resources. Moreover, import protection is likely to be met by foreign retaliation that reduces employment in the more efficient export industries.

4. For a discussion of social benefit-cost accounting, see I. M. D. Little and J. A. Mirrlees, *Project Appraisal and Planning for Developing Countries* (New York: Basic Books, 1974); see also I. M. D. Little, *Economic Development, Theory, Policy and International Relations,* a Twentieth Century Fund book (New York: Basic Books, 1982), pp. 127–29.

5. Social benefit-cost accounting is by definition a quantitative and not simply a normative tool. It does require assigning values to benefits and costs that do not have a market value, but in doing so an effort is made to approximate what the market value would be if the utilities or disutilities were actually traded.

Chapter 2

1. It might be objected that since the earth itself is finite, there must always be some material limit on growth. However, it is conceivable that in several hundred

years science will have the capability of utilizing the material resources of the entire solar system and beyond.

2. Donald A. Brobst, "Fundamental Concepts for the Analysis of Resource Availability," in *Scarcity and Growth Reconsidered,* ed. V. Kerry Smith (Baltimore, Md.: Johns Hopkins University Press for Resources for the Future, 1979), p. 123.

3. Ibid. It is estimated there are 83 kg of aluminum and 48 kg of iron per mt of continental crust. However, there are only 50 gms of copper and 0.0035 gms of gold per mt in the earth's crust.

4. Ibid., pp. 123–27; B. J. Skinner, "A Second Iron-Age Ahead?" *American Scientist* 64 (1976): 258–69.

5. Brobst, "Fundamental Concepts," pp. 125–26. This hypothesis was first suggested by V. E. McKelvey, "Relation of Reserves of the Elements to Their Crustal Abundance," *American Journal of Science* 258-A (1960): 234–41; *Mineral Resources and the Environment,* Report by the Committee on Mineral Resources and the Environment (National Research Council) (Washington, D.C.: National Academy of Sciences, 1975).

6. Brobst, "Fundamental Concepts," pp. 127–29.

7. For a list of USGS estimates of potentially recoverable mineral resources, see D. A. Brobst and W. P. Pratt, eds., *United States Mineral Resources,* Geological Survey Professional Paper 820 (Washington, D.C.: U.S. Government Printing Office, 1973); for a discussion of the relationship between these figures and potential reserves, see F. E. Trainer, "Potentially Recoverable Resources: How Recoverable?" *Resources Policy,* March 1982, pp. 41–52.

8. Bureau of Mines, *Mineral Commodity Summaries 1983* (Washington, D.C.: U.S. Department of the Interior, 1983), pp. 98–99.

9. The tendency of the real cost of a mineral to equal its real price may be explained as follows: The real unit cost of production does not include the rental value of resources (i.e., the amounts paid to owners of resources for their scarcity value). Given a fixed supply of resources, owners will earn higher rents as demand increases regardless of the real cost of extraction. Thus, the owner of a large petroleum field will be able to demand higher prices (and, hence, higher rents) for the petroleum even though there is no change in the cost of producing oil. On the other hand, if resources are not fixed in amount (as is the case with virtually all nonfuel minerals) but can be expanded with additional outlays for exploration and development, output will tend to increase up to the point where the real cost of finding and producing additional resources is equal to the real price. The owners of resources that have been found at lower cost in the past would enjoy larger rents as prices rise, but the new higher cost output will yield no rents if costs are equal to prices.

10. H. G. Barnett and Chandler Morse, *Scarcity and Growth: The Economics of Natural Resource Availability* (Baltimore, Md.: Johns Hopkins University Press for Resources for the Future, 1963); Barnett, "Scarcity and Growth Revisited," in *Scarcity and Growth Reconsidered,* ed. Smith, pp. 163–217.

11. Ibid., pp. 165–74.

12. Ibid., pp. 174–76.

13. Orris C. Herfindahl, *Copper Costs and Prices: 1870–1957* (Baltimore, Md.: Johns Hopkins University Press for Resources for the Future, 1959), pp. 208–9.

14. Raymond F. Mikesell, *The World Copper Industry* (Baltimore, Md.: Johns Hopkins University Press for Resources for the Future, 1979), pp. 126–29. There is considerable evidence that the real cost of producing copper in 1957 was maintained until the early 1970s. However, there is also evidence that after 1973 (at least until 1977) real costs rose significantly. This was mainly accounted for by factors, such as increasing labor and capital costs, other than a decline in average grades mined in the world.

15. Gardner M. Brown, Jr., and Barry Field, "The Adequacy of Measures of Signaling Scarcity of Natural Resources," and Anthony C. Fisher, "Measures of Natural Resource Scarcity," in *Scarcity and Growth Reconsidered,* ed. Smith, pp. 218–20 and 249–73.

16. For an extreme view that denies the existence of scarcity of any natural resource, see Julian L. Simon, *The Ultimate Resource* (Princeton, N.J.: Princeton University Press, 1981), chaps. 2 and 3. Simon assumes virtually unlimited substitutability of abundant for scarce resources, including petroleum.

17. This is one of the reasons why mineral prices tend to fall whenever interest rates rise.

18. The reserve base in table 2-1 is somewhat larger than economic reserves reported by the Bureau of Mines since it includes both marginally economic reserves and some currently subeconomic resources that have a reasonable potential for becoming economical over the next two decades and at real prices somewhat higher than current levels. The reserve base in table 2-1 does not include hypothetical and speculative (undiscovered) resources, which in most cases are many times larger than the reserve base. Estimates of undiscovered resources are included as world resources of selected minerals given in table 2-2.

19. See table 2-4.

20. For Bureau of Mines projected consumption rates of growth for 1978–2000, see table 2-1; for historical rates of production for 1947–74, see Hans H. Landsberg and J. E. Tilton, "Nonfuel Minerals," in *Current Issues in Natural Resource Policy,* ed. P. R. Portney (Baltimore, Md.: Johns Hopkins University Press for Resources for the Future, 1982), chap. 3.

21. Wassily Leontief, James C. M. Koo, Sylvia Nasar, and Ira Sohn, *The Future of Nonfuel Minerals in the U.S. and World Economy* (Lexington, Mass.: Lexington Books, 1983).

22. Ibid., pp. 39–64. Changes in technology for the final product sectors consuming each of the twenty-six nonfuel minerals from 1972 to 2000 are projected on the basis of present knowledge. The production coefficients for the years 2000 to 2030 are assumed to be the same as those in the year 2000.

23. See ibid., p. 217, for GDP growth-rate projections by region. For example, projected average annual growth rates for medium-income Latin American countries are 3.7 and 2.8 percent for the optimistic and pessimistic scenarios, respectively, for the 2000–2030 period.

24. Ibid. Estimates on world resources are given on p. 173.

25. Ibid., chap. 8.

26. This statement implies the absence of serious shortages as measured by trends in real prices.

27. World Bank, *World Development Report 1983* (New York: Oxford University Press, 1983), app. table 19, p. 184.

28. World Bank, *World Development Report 1982* (New York: Oxford University Press, 1982), pp. 21–22.

29. World Bank, *World Development Report 1984* (New York: Oxford University Press, 1984), p. 7. The population of developed countries, including the Soviet bloc (USSR plus its European satellites), is projected to rise from 1.2 billion in 1980 to 1.4 billion in 2050, while the population of developing countries would grow from over 3.6 billion to 8.4 billion.

30. More than 8 percent of the earth's crust consists of aluminum, the most abundant mineral on earth; iron takes second place at 5 percent.

31. Wolfgang Gluschke, Joseph Shaw, and Bension Varon, *Copper: The Next Fifteen Years* (Boston, Mass.: Reidel Publishing, 1979), p. 51.

32. H. L. Martin and L. S. Jen, "Are Ore-Grades Declining—The Canadian Experience, 1939–1989" (Paper presented at the Mineral Exploration Seminar sponsored by the International Institute for Applied Systems Analysis, Laxenburg, Austria, December 1983). For example, they state that the large porphyry copper operations were not started as a consequence of depletion of high-grade ores, but rather as a consequence of a shift to large-scale open pit mining of copper porphyry ore bodies as an alternative to the much more expensive underground mining of higher-grade copper deposits. Hence, the average porphyry copper grade mine in Canada shows a decline in time, which is related to the growth of the size of operations rather than to a decline in ore grades. The authors conclude that "it appears that the effects of resource depletion on grades mined come into play at a slower rate than is generally assumed."

33. Donald A. Cranstone, "Canadian Mineral Discovery Experience since World War II" (Paper presented at the Mineral Exploration Seminar sponsored by the International Institute for Applied Systems Analysis, Laxenburg, Austria, December 1983).

34. Marian Radetzki, "Long-Run Price Prospects for Aluminum and Copper," *Natural Resources Forum* 7, no. 1 (January 1983): 33. The author concludes that the fall in ore grades in the world copper industry over the remainder of the present century "will be very limited, and that upward cost-push will be easy to compensate by cost-reducing technological progress."

35. Gerald Pollio, "The Outlook for Major Metals through the Year 2000: An Updated View," *Journal of Resource Management and Technology* 12, no. 2 (April 1983): 115.

36. See, for example, Anthony C. Fisher, *Resource and Environmental Economics* (Cambridge: Cambridge University Press, 1981), chap. 6.

37. John E. Tilton, "The Nature and Significance of Materials Substitution," in *Materials Substitution: Experience of Tin-Using Industries*, mimeographed (University Park, Pa.: Pennsylvania State University, 1980), chap. 5.

38. Committee on Mineral Technology of the National Academy of Sciences, *Technology Innovation and Forces for Change in the Mineral Industry* (Washington, D.C.: National Academy of Sciences, 1978), p. 22.

39. K. M. Clark and Z. Grilliches, *Productivity, Growth and R&D at the Business Level: Results from the PIMS Data Base*, Working Paper 916 (Cambridge, Mass.: National Bureau of Economic Research, June 1982), p. 9.

40. Committee on Mineral Technology, *Technology Innovation,* chap. 2.

41. E. S. Bonczar and J. E. Tilton, *An Economic Analysis of the Determinants of Metal Recycling in the United States: A Case Study of Secondary Copper,* Final Report to the Bureau of Mines (Washington, D.C.: U.S. Department of the Interior, 1975), p. 2; Mikesell, *World Copper Industry,* pp. 339–49.

42. Bureau of Mines, *Mineral Facts and Problems 1980* (Washington, D.C.: U.S. Government Printing Office, 1981), pp. 460–61.

43. Bureau of Mines, *Mineral Commodity Summaries 1982* (Washington, D.C.: U.S. Department of the Interior, 1983), p. 84.

44. Old scrap is material from discarded products, while new scrap is generated from metal fabricating operations. New scrap does not constitute a net addition to supplies of the material and its amount is largely determined by the level of current consumption. Old scrap from discarded materials varies greatly in the percentage of metal in the materials and in the cost and technical capability of refining it or using it directly in production.

Chapter 3

1. General Agreement on Tariffs and Trade, *International Trade, 1982/83* (Geneva, 1983), app. table A-17.

2. Bureau of Mines, *Mineral Commodity Summaries 1982* (Washington, D.C.: U.S. Department of the Interior, 1983), p. 94. Data not adjusted for differences in metallic content of ores and ferroalloys.

3. Alumina is processed from bauxite ore and is the primary material for the production of aluminum.

4. In this study South Africa is classified as a developed but not an industrial country. This is also true in the case of World Bank reports, although United Nations documents usually include South Africa in the developing country category.

5. For data on Soviet exports of manganese and chromite ore, see Walter C. Labys, "An Economic Analysis of the Stability of East-West Mineral Trade Patterns" (Paper presented at the Conference on East-West Mineral Trade sponsored by the International Institute for Applied Systems Analysis [IIASA], Vienna, Austria, March 1984).

6. V. B. Strishkov, "The Mineral Industry of the U.S.S.R.," in Bureau of Mines, *Minerals Yearbook 1982,* vol. 3 (Washington, D.C.: U.S. Government Printing Office, 1982), pp. 1003–74.

7. Marian Radetzki, "China's Foreign Trade in Metals Minerals—Performance and Prospects" (Paper presented at the Conference on East-West Mineral Trade sponsored by the International Institute for Applied Systems Analysis, Vienna, Austria, March 1984).

8. Leonard L. Fischman, *World Mineral Trends and U.S. Supply Problems* (Baltimore, Md.: Johns Hopkins University Press for Resources for the Future, 1980), chap. 9.

9. Full economic costs include operating costs, taxes, interest on borrowed capital, depreciation of fixed assets, depletion of ore body, and after-tax profits sufficient to attract a similar investment. To reflect full *real* economic costs, fixed capital costs should include the current cost of constructing a new mining complex with the

same capacity as representative mines already in operation. However, the increased productivity of a new mine might offset any increase in the real cost of a new mine over the real cost (after adjustment for depreciation) of older mines. Also, operating costs (in constant prices) might be lower for new mines than for older ones.

10. The World Bank price projections are based on optimistic projections of increases in developing country production of nonfuel minerals that may not be realized if these countries are unable to finance additional capital requirements.

11. Raymond F. Mikesell, *New Patterns of World Mineral Development* (Washington, D.C.: National Planning Association, September 1979), pp. 12–13 and app. D. In an earlier study I estimated capital requirements for nonfuel minerals (excluding steel capacity) over the period 1977–2000 at about $300 billion, or $12.5 billion annually in 1977 dollars, of which $5 billion would represent the share for developing countries. Since that study was completed, projected rates of growth of consumption of nonfuel minerals have been substantially reduced.

12. World Bank, *The Outlook for Primary Commodities,* Staff Commodity Working Paper no. 9 (Washington, D.C.: World Bank, 1983); Chemical Bank of New York estimates are given in Gerald Pollio, "The Outlook for Major Metals through the Year 2000: An Updated View," *Journal of Resource Management and Technology* 12, no. 2 (April 1983): 113–18.

13. The World Bank study was first circulated as an internal document in July, 1982, on the basis of the Bank's consumption growth estimates made during the first half of 1982. Since then the World Bank's own published projections of rates of growth in consumption have been reduced somewhat, but are still higher than those used by Chemical Bank in its study published in 1983.

14. These estimates do not allow for any rise in the real cost of mineral capacities over the period of the projection.

15. A portion of the local currency outlays would give rise to indirect foreign exchange requirements.

16. World Bank, *Outlook for Primary Commodities,* pp. 97–100.

17. Pollio, "Outlook for Major Metals," p. 114.

18. Paul E. Sigmund, *Multinationals in Latin America: The Politics of Nationalization,* Twentieth Century Fund Study (Madison, Wis.: University of Wisconsin Press, 1980); J. Frederick Truitt, *Expropriation of Private Investment* (Bloomington, Ind.: Indiana University Graduate School of Business, 1974); Raymond F. Mikesell, *Foreign Investment in Petroleum and Mineral Industries* (Baltimore, Md.: Johns Hopkins University Press for Resources for the Future, 1971).

19. Department of Commerce, *U.S. Business Investments in Foreign Countries* (Washington, D.C.: U.S. Government Printing Office, 1960), p. 91; and *Survey of Current Business* (Department of Commerce, Washington, D.C.), October 1971, p. 28. Book value includes equity interest of U.S. corporations plus long-term debt and other liabilities of foreign subsidiaries to the U.S. parent company. Bank loans and other indebtedness of the foreign affiliates to third parties are not included as direct investments.

20. Ibid., p. 90, and *Survey of Current Business,* August 1982, p. 22.

21. Organization for Economic Cooperation and Development, *Investing in Developing Countries* (Paris: Organization for Economic Cooperation and Development, November 1982), p. 24.

22. *Survey of Current Business,* October 1970, pp. 28–29, and August 1982, p. 22.

23. The vast bulk of the U.S. direct foreign mineral investment in developing countries is in nonfuel minerals, while a portion of U.S. direct foreign investment in developed countries is in coal and uranium.

24. Philip Crowson, "Investment and Future Mineral Production," *CIPEC Quarterly Review,* April-June 1982, p. 52. This data includes coal, but even if coal were eliminated the percentage of total annual mineral investment in developing countries declined between 1970 and 1980.

25. For a review and analysis of Australian and Canadian foreign investment screening systems, see David L. Anderson, *Foreign Investment Control in the Canadian Mineral Sector: Lessons from Australian Experience* (Kingston, Ont.: Centre for Resource Studies, Queens University, 1984), chap. 7.

26. See Raymond F. Mikesell, *Petroleum Company Operations and Agreements in the Developing Countries* (Baltimore, Md.: Johns Hopkins University Press for Resources for the Future, 1984), chap. 4 for discussion of the role of risk in investment decisions.

27. For an analysis of modern mining contracts, see Raymond F. Mikesell, *Foreign Investment in Mining Projects* (Boston, Mass.: Oelgeschlager, Gunn and Hain for Fund for Multinational Management Education, 1983).

28. For an analysis of the effects of various fiscal arrangements on the expected rates of return on mineral projects with different degrees of risk, see Mikesell, *Petroleum Company Operations,* chap. 4, and Ross Garnaut and Anthony Clunies Ross, *Taxation of Mineral Rents* (Oxford: Clarendon Press, 1983), chap. 9.

29. Investors usually employ the "internal rate of return" rather than the ratio of after-tax net profits to invested capital (the accounting rate of return) in calculating the expected rate of return on an investment. The internal rate of return takes into account the time pattern of net revenues and costs. The internal rate of return is defined as that rate of discount (interest) that equates the present (discounted) value of the revenue stream with the present value of the cash outlay.

30. For a discussion of the negotiation of the Ok Tedi mining agreement, see William S. Pintz, *Ok Tedi: Evaluation of a Third World Mining Project* (London: Mining Journal Books, 1984).

31. For a discussion of the national equity participation provisions of PNG, Botswana, and Indonesian mining agreements, see Mikesell, *Foreign Investment in Mining Projects,* chaps. 3, 5, 6, and 13.

32. This was less true in the case of the 20 percent equity interest acquired by the PNG government before actual mine construction began.

33. This regional workshop in which the author participated as an expert on mining contracts was sponsored by the Indonesian Mining Association and held in Jakarta, Indonesia, September, 1983. Countries represented included Indonesia, the Philippines, Malaysia, and PNG.

34. Marian Radetzki, *State Mining Enterprises: An Investigation into Their Impact on International Mineral Markets* (Washington, D.C.: Resources for the Future, 1985), chap. 3.

35. Ibid.; see also Raymond Vernon and Brian Levy, "State-Owned Enterprises in the World Economy: The Case of Iron Ore," *Public Enterprises in Less*

Developed Countries, ed. Leroy P. Jones (Cambridge: Cambridge University Press, 1982), chap. 9.

36. This has tended to be the case in some mineral countries, including Venezuela and Zambia, but not in Chile and Peru, although both of the latter countries have nationalized large segments of the foreign-owned mining industry.

37. When Kennecott Copper Corporation's large El Teniente mine in Chile was nationalized, there were only five Americans working at the mine.

38. The elimination of foreign equity investment to the maximum extent possible has been a major theme of a series of reports prepared within the United Nations for the Committee on Natural Resources of the Economic and Social Council under the title of "Permanent Sovereignty Over Natural Resources." The first of these reports was issued in 1953 and a later report was published under this same title, no. E/C.7/1983/5 (New York: United Nations, April 1983).

39. The loan agreement with the creditors usually provides that proceeds from export of products be placed in a trust account with a foreign bank from which account payments are made to the creditors.

40. For examples of the operation of completion guarantees under these circumstances, see Mikesell, *Foreign Investment in Mining,* chaps. 3, 6, and 11.

41. J. F. Joklik, "Upstream Metals Mining" (Paper presented at a Sohio Securities Analysts Meeting, 1983), p. 96. A recent study by Kennecott Minerals Corporation estimates that the 100 percent government-owned copper mines in Chile will require about $6.5 billion between 1982 and 2000 in order to replace depleted ore bodies and increase copper production by 20 percent. If we apply this estimate of financial requirements to the 2.5 million mt per year of copper capacity (1981) of the state enterprises in developing countries with an equity interest in excess of 50 percent, the annual financial requirements for copper alone will be in excess of $1 billion. CODELCO's refined copper capacity is about 900,000 mt per year.

42. See, for example, Charles F. Barber, "Mines Are Not Public Works Projects," *American Mining Congress Journal,* June 8, 1983, p. 9; see also Paul W. MacAvoy, "A Policy That Closes U.S. Copper Mines," *New York Times,* December 19, 1982, p. 3; on alleged World Bank and International Monetary Fund complicity in artificially low mineral prices, see "International Lending Policies and Their Effects on Minerals and Metals Industry," American Mining Congress position paper, *American Mining Congress Journal,* July 7, 1983, pp. 10–11.

43. William G. Siedenburg, *Copper Quarterly* (New York: Smith Barney, Harris Upham and Co., May 1983), pp. 21–22.

44. Private mining firms in developing countries may also be constrained from reducing output or temporarily shutting down when operating costs exceed revenues because of employment guarantees and large labor termination payments. In addition, the host government may make it difficult for domestic private and foreign-owned mines to curtail production. They are sometimes threatened with expropriation if they do so.

45. U.S. Congress, *Trade Act of 1974, Title III: Unfair Trade Practice Relief,* 93d Cong., 2d sess., 1974. The 1974 Trade Act gives the U.S. government such retaliatory authority.

46. Mineracao Rio do Norte, the largest producer and exporter of bauxite in

Brazil, is owned by CVRD (46 percent), Alcan (24 percent), and Brasiliero de Aluminion (10 percent).

47. Cia de Mineracao Serra Gerol is the name of the Brazilian iron ore joint venture.

48. Vernon and Levy, "State-Owned Enterprises," p. 184.

Chapter 4

1. Once mines are abandoned it is very costly and time-consuming to restore them to productive operation, if that ever happens. The mines may become flooded and the equipment allowed to deteriorate or be sold as scrap. In addition, the labor force may have to be reconstituted and retrained.

2. By definition, "pure competition" exists when sellers sell without taking account of the effect of their sales on the market price; buyers buy at the cheapest price without regard to the effect of their purchases on the market. Excluding transportation costs, there is a single market price at any particular time. If there is more than one market for the commodity, the prices are the same in all markets as a consequence of arbitrage between markets. Pure competition rarely if ever exists, but markets *are* characterized by varying degrees of competition, depending upon the ability of sellers or buyers to influence prices, or the effect of government controls.

3. In an oligopolistic market, producers take into account the impact of their own sales and prices on the sales and prices of their competitors. Changes in price often take place when a leading firm changes its price, followed by other producers. Under these circumstances, prices are not set by overt collusion. Overt collusion among producers means monopoly price setting, which, of course, violates antitrust laws in the United States and some other Western countries. Oligopolistic markets may be fairly competitive without collusion. Competition is more likely on the basis of product quality and marketing than through open price competition. When demand falls, producers may offer secret discounts to customers rather than reduce their published prices.

4. The long-term equilibrium price of a commodity is equal to the full cost of production, including profits sufficient to attract new capital. The tendency for prices to equal full costs of production over the longer run requires more-or-less free entry into the industry. This is the case for most internationally traded minerals whose production is widely diversified among different companies in a number of countries. There may be relatively free entry even though the market is not highly competitive. However, free entry would not exist if almost all of a mineral were controlled by one or two firms.

5. Commodities traded on the LME are aluminum, copper, gold, lead, nickel, silver, tin, and zinc. Commodities traded on the COMEX are aluminum, copper, gold, and silver. For a description of metal exchange operations, see Peter Robbins, *Guide to Nonferrous Metals and Their Markets* (New York: Kogan Page, 1982), pp. 33–34 and app. 5.

6. See table 2-6. In my opinion the World Bank staff price projections shown in this table are too low. They are, in fact, markedly lower than those published by the World Bank in 1982; also see World Bank, *The Outlook for Primary Commodities,*

Staff Commodity Working Paper no. 11 (Washington, D.C.: World Bank, 1982), p. 32.

7. Blister copper is smelted but not refined. Refining removes impurities to produce over 99 percent pure copper.

8. Bauxite is refined into alumina at a ratio of about four-and-one-half tons of dried bauxite to two tons of alumina. Two tons of alumina will produce one ton of aluminum metal.

9. U.S. Department of Interior, *Aluminum: Mineral Commodity Profiles* (Washington, D.C.: U.S. Government Printing Office, May 1978), pp. 29, 35, and 37.

10. United Nations Conference on Trade and Development (UNCTAD) Secretariat, *Processing and Marketing of Bauxite/Alumina/Aluminum: Areas for International Cooperation* (Geneva: United Nations, 1981), table 15.

11. Ibid., table 11.

12. Ibid., tables 9 and 10.

13. Ibid., chap. 4.

14. Robbins, *Guide to Nonferrous Metals,* p. 49.

15. UNCTAD, *Processing and Marketing Bauxite,* pp. 53–57.

16. U.S. Department of Interior, *Iron Ore: Mineral Commodity Profiles* (Washington, D.C.: U.S. Government Printing Office, May 1978), pp. 2–3.

17. See Walter C. Labys, *Market Structure, Bargaining Power and Resource Price Formation* (Lexington, Mass.: D. C. Heath, 1980), pp. 189–94, for a discussion of iron ore price formation.

18. A brief description of the U.S. iron and steel industries is found in U.S. Department of Interior, *Iron and Steel: Mineral Commodity Profiles* (Washington, D.C.: U.S. Government Printing Office, July 1978), pp. 3–14.

19. For a description of marketing practices of U.S. producers and importers to the U.S. market, see U.S. International Trade Commission, *Conditions of Competition in the Western U.S. Steel Market between Certain Domestic and Foreign Steel Products* (Washington, D.C.: U.S. International Trade Commission, March 1979), pp. 48–63.

20. UNCTAD, *The Iron and Steel Industry* (New York: United Nations, 1978), chap. 8.

21. U.S. International Trade Commission, *Survey and Analysis of Government Ownership of Market Economy Countries: A Study of Steel, Automobiles and Iron Ore* (Washington, D.C.: U.S. International Trade Commission, May 1978), p. 174.

22. Wirebars are usually 4⅝ inches wide and up to 4½ inches thick and weigh 200 to 300 pounds. Cathode copper is produced by electrolytic refining. CCR measures about ¼ inch in diameter and is produced in long coils weighing up to 15,000 pounds.

23. Alloyed copper products include various types of bronze and brass.

24. CODELCO in Chile; Gecamines in Zaire; and Zimco in Zambia.

25. Primary smelting and refining capacity use copper ores as the major input, while secondary smelters and refineries use scrap as the major input.

26. For a discussion of the rationale for the U.S. producer-price system, see Raymond F. Mikesell, *The World Copper Industry* (Baltimore, Md.: Johns Hopkins University Press for Resources for the Future, 1979), pp. 111–16.

27. For a description of the current zinc industry, see S. Gupta, *The World's Zinc Industry* (Lexington, Mass.: D.C. Heath, 1982), p. 50.

28. Robbins, *Guide to Nonferrous Metals,* pp. 154–55; Gupta, *World's Zinc,* pp. 24–31.

29. For a description of commercial forms and various uses of nickel, see N. A. Matthews and S. F. Sibley, "Nickel," in Bureau of Mines, *Mineral Facts and Problems 1980* (Washington, D.C.: U.S. Government Printing Office, 1981), pp. 611–15.

30. On pricing methods, see Robbins, *Guide to Nonferrous Metals,* pp. 117–18.

31. Ibid., pp. 142–45.

32. UNCTAD, *The Processing and Marketing of Manganese: Areas for International Cooperation* (Geneva: United Nations, 1981), table 7.

33. Robbins, *Guide to Nonferrous Metals,* pp. 116–18.

34. Chromite ore typically contains 22 to 38 percent chromium. Ferrochromium is a metal alloy containing up to 70 percent chromium.

35. Robbins, *Guide to Nonferrous Metals,* pp. 85–86; see also Bureau of Mines, *Mineral Commodity Summaries 1982* (Washington, D.C.: Department of Interior, 1983), pp. 32–33.

36. Bureau of Mines, *Mineral Commodity Summaries 1983* (Washington, D.C.: Department of Interior, 1984), pp. 36–37; J. T. Kummer, "Cobalt," in Bureau of Mines, *Minerals Yearbook 1980* (Washington, D.C.: U.S. Government Printing Office, 1981), pp. 237–46.

37. S. F. Sibley, "Cobalt," in Bureau of Mines, *Mineral Facts and Problems 1980* (Washington, D.C.: U.S. Government Printing Office), p. 210.

38. Platinum and palladium are traded on the New York Mercentile Exchange.

Chapter 5

1. For example, the ad valorem duty on copper is 1 percent; structural steel, 0.9 percent; carbon steel, 6 percent (average); zinc, 1.5 percent (after 1986); and aluminum will pay no duty after 1986. Iron ore, bauxite, and alumina are now duty free.

2. The Nixon administration rejected the long conventional war scenario; in April, 1973, Nixon announced new national defense stockpile guidelines that would provide materials needed to cover only the first year of a major conflict in Europe and Asia instead of the previous guideline assuming a three-year war. In the course of hearings on the national defense stockpile, Senator Proxmire said that "Reliance on the World War II scenario borders on the absurd. Why are we stockpiling for this age-worn contingency when logic points us in another direction?" (U.S. Congress, Senate, Committee on Banking, Housing and Urban Affairs, *Hearings, Strategic Stockpile Policy,* 95th Cong., 2d sess., November 14, 1978, pp. 1–3). The World War II assumption has been widely criticized. For a discussion of this, see my monograph, *National Defense Stockpile: Historical Review and Current Assessment* (Washington, D.C.: American Enterprise Institute, 1986).

3. "Agency Decision on Copper-Import Curbs Could Recast Shape of Sagging Industry," *Wall Street Journal,* June 11, 1984, p. 32. This statement was attributed to

Richard de J. Osborne, president of ASARCO, but similar statements have been made by other representatives of the copper industry.

4. Fifty-five percent would come from Canada and Mexico plus secondary production, and 30 percent would come from U.S. primary production—60 percent of 1984's 50 percent of requirements.

5. William G. Siedenburg, *Copper Quarterly* (New York: Smith Barney, Harris Upham and Co., June 10, 1985), pp. 18–22.

6. John E. Tilton, "Comparative Advantage in Mining" (Working paper, International Institute for Applied Systems Analysis, Laxenburg, Austria, September 1983). The author found a significant positive statistical relationship between the growth of mineral production of major producing countries and the growth of their mineral reserves for bauxite, copper, iron ore, tin, and zinc over the period 1950–80.

7. William G. Siedenburg, *Copper Quarterly* (New York: Smith Barney, Harris Upham and Co., October 1983), pp. 16–17.

8. World Bank, *Price Prospects for Major Primary Commodities,* vol. 4 (Washington, D.C.: World Bank, September 1984), p. 22. In September, 1984, the World Bank staff forecast an 18 percent rise in the world price of copper (in constant prices) between 1983 and 1990.

9. A. F. Barsotti and R. D. Rosenkranz, "Estimated Costs for the Recovery of Copper from Demonstrated Resources in Market Economy Countries," *Natural Resources Forum,* April 1983, p. 108.

10. Frederick R. Demler, *Copper Quarterly Report* (New York: Drexel Burnham Lambert, August 1984), pp. 52–53.

11. Moderately optimistic forecasts of rising U.S. copper production were made in Demler, *Copper Quarterly Report,* pp. 36–38, and William G. Siedenburg, *Copper Quarterly* (New York: Smith Barney, Harris Upham and Co., October 1983), pp. 17–20.

12. Statistics through 1982 were taken from Bureau of Mines, *Mineral Commodity Summaries 1983* (Washington, D.C.: U.S. Department of Interior, 1984), p. 78; 1984 data on net import reliance are not reported on a comparable basis with earlier figures.

13. U.S. Department of Commerce, *Steel: Supply and Demand in the 1980s* (Washington, D.C.: U.S. Government Printing Office, July 1982), p. 16.

14. "U.S. Steel to Cut 15,430 Jobs: Three Plants to Be Closed in April," *New York Times,* December 28, 1983, p. 1.

15. "LTV-Republic Merger Plan Is Approved," *Wall Street Journal,* May 21, 1984, p. 14.

16. Donald E. Barnett, "A Restructured Steel Industry," *New York Times,* February 2, 1984, p. 29; "The Rebirth of Steel," *Wall Street Journal,* February 16, 1984, p. 28.

17. Minimills produce steel in electric furnaces.

18. Ingo Walter, "Structural Adjustment and Trade Policy in the International Steel Industry," in *Trade Policy in the 1980s,* ed. William R. Cline (Cambridge, Mass.: MIT Press, 1983), p. 510.

19. "Rebirth of Steel," p. 28.

20. World Bank, *Price Prospects for Major Primary Commodities,* vol. 4 (Washington, D.C.: World Bank, September 1984), pp. 119–22.

21. John E. Tilton, "Import Dependence and the Future Availability of Aluminum in the United States," in National Materials Advisory Board, *Assessment of Selected Materials Issues* (Washington, D.C.: National Academy Press, 1981), pp. 85–91.

22. Ibid.

23. See chap. 7 for a discussion of vulnerability to bauxite and alumina supply disruption.

24. The United States buys steel from Argentina, Brazil, South Korea, and Taiwan; over half of its copper imports come from developing countries.

25. Economists generally prefer subsidies to import restrictions for two reasons. First, import restrictions raise prices of restricted imports above world levels, thereby increasing the costs and impairing the competitive position of industries using these commodities as inputs. Second, purchasers of the restricted commodities are taxed by the amount of the tariff (or quota equivalent to the tariff) in order to achieve an alleged benefit to the economy as a whole. If assistance to the domestic industry benefits the economy as a whole, that assistance should be paid for out of general revenues in the form of a subsidy rather than taxing the purchasers of commodities in question.

26. Robert W. Crandall, *The U.S. Steel Industry in Recurrent Crisis* (Washington, D.C.: Brookings Institution, 1981), p. 103.

27. VRAs are employed in lieu of direct import restrictions in cases of dumping, of alleged unfair trade practices by foreign exporters, or of a judgment by the International Trade Commission under the Trade Act of 1974 that imports significantly harm a U.S. industry. Foreign acceptance of VRAs arises from the fear that direct U.S. action would be more restrictive than a "voluntary" limitation. The United States prefers VRAs because the administration is reluctant to take measures that might violate the General Agreement on Tariffs and Trade, or that might trigger retaliatory limitations on U.S. exports. VRAs are a form of international commercial blackmail; they have been used in such product categories as automobiles, steel, and textiles.

28. The two principal unfair trade practices are dumping, defined as selling abroad at less than the price charged in the exporting country (or according to the Trade Act of 1974, below the cost of production), and the subsidization of production or export. In both cases complaints are filed with the Department of Commerce. If the department's finding bears out the complaint, special antidumping tariffs or countervailing duties (in the case of subsidies) may be levied against products from the offending countries. However, the ITC first has to find that such imports cause or threaten to cause serious injury to U.S. competitors. Even without charges of unfair trade practices, any U.S. company or labor union may appeal to the ITC for import relief on grounds that they are a major cause of serious injury. If the ITC then makes a finding of injury, it must propose relief to the president, who may accept or reject the recommendation, or provide an alternative measure, such as trade adjustment assistance or a VRA.

29. "Europe Agrees to Cut Steel Exports to U.S. As Ten-Month Standoff Ends," *Wall Street Journal,* August 22, 1982, p. 3; see also "How the Steel Industry Dictated the Trade Agreement," *Wall Street Journal,* November 23, 1982, p. 28.

30. According to U.S. trade law, the substitution of a VRA for an action against unfair trade practices requires agreement by the firms filing the complaints.

31. The Bethlehem–United Steelworkers petition was not joined or strongly supported by other large integrated steel companies because they were parties to the VRA between the U.S. government and the EEC.

32. International Trade Commission, *Carbon and Certain Alloy Steel Products,* Report to the President on Investigation No. TA-201-51 under Section 201 of Trade Act of 1974 (Washington, D.C.: International Trade Commission, July 1984), 1:1–5; Commissioners A. E. Eckes, S. G. Lodwick, and D. B. Rohr constituted the majority in their finding of serious injury: they recommended the imposition of tariffs and quotas on imported steel products. Commissioners P. Stern (chair) and S. W. Liebeler (vice-chair) dissented.

33. Ibid., pp. 88–103.

34. Ibid., p. 119.

35. Ibid., p. 156.

36. "ITC Proposes Quotas, Tariffs, on Steel Imports," *Wall Street Journal,* July 12, 1984, p. 3. Commissioners Stern and Liebeler, voting against the imposition of import restrictions that were announced in the Bethlehem case June 12, 1984, argued that import restrictions would "institutionalize inefficiency" in the domestic steel industry.

37. "Voluntary Import Restraint: Effects Similar to Quotas," *New York Times,* September 20, 1984, p. 27; see also, "U.S. Steelmakers Buoyed by Reagan Plan to Seek Voluntary Restraints on Imports," *Wall Street Journal,* September 20, 1984, p. 2.

38. "Reagan Faces Flap with Europeans over Proposed New Steel-Import Curbs," *Wall Street Journal,* June 28, 1984, p. 29; "Common Market Retaliating, Curbs Imports from U.S.," *New York Times,* January 14, 1984, p. 26.

39. "National in Japanese Steel Deal: Nippon Kokan Will Buy Half of U.S. Entity," *New York Times,* April 25, 1984, p. 25; "Kawasaki Steel to Acquire 25 Percent of U.S. Venture: Plan Is to Buy Factory Idled by Kaiser in California," *Wall Street Journal,* July 17, 1984, p. 2.

40. In April, 1983, Congressman Morris K. Udall of Arizona (Arizona produces 67 percent of the nation's copper) introduced a bill to impose a duty on imported copper equal to the cost advantage enjoyed by foreign producers not subject to environmental regulations comparable to those in the United States. See Udall's statement in *Congressional Record,* April 4, 1983 (reprinted in *American Mining Congress Journal,* April 13, 1983, p. 14).

41. International Trade Commission, *Unwrought Copper,* Report to the President on Investigation No. TA-201-52 under Section 201 of Trade Act of 1974 (Washington, D.C.: International Trade Commission, July 1984), pp. 1–2. Commissioners Stern and Rohr recommended the 5 percent per pound increase in duty; Commissioners Eckes and Lodwick recommended quantitative restrictions; and Commissioner Liebeler recommended no import restrictions.

42. Ibid., pp. 31–43.

43. Ibid., pp. 47–52.

44. "Steel Firms Had Strong Quarter: Weakening Seen," *Wall Street Journal,* July 5, 1984, p. 4; "U.S. Steel Posts Second-Quarter Net of $140 Million," *Wall Street Journal,* August 1, 1984, p. 2.

45. P. C. F. Crowson, "Investment and Future Mineral Production," *CIPEC Quarterly Review,* April-June 1982, pp. 43–44.

46. C. G. Streets, "The Cost of Primary Copper in December 1978," mimeographed (London: Consolidated Gold Fields, Ltd., July 1979), p. 5.

47. Ibid., pp. 8–11.

48. Donald F. Barnett and Lewis Schorsch, *Steel: Upheaval in a Basic Industry* (Cambridge, Mass.: Ballinger, 1983), p. 23.

49. Department of Commerce, *Steel in the 1980s,* pp. 9, 23–25.

50. Barnett and Schorsch, *Steel: Upheaval,* p. 23.

51. Robert W. Crandall, "The Economics of the Current Steel Crisis in OECD Member Countries," *Steel in the Eighties* (Paris: Organization for Economic Cooperation and Development, 1980), p. 143.

52. Total Latin American and Asian (excluding Japan) output was nearly 44 million mt in 1978.

53. The U.S. integrated steel industry represented by large steel firms, such as U.S. Steel and Bethlehem, produces steel from iron ore using coke ovens, blast furnaces, open-hearth furnaces, or basic oxygen furnaces. About one-fourth of U.S. steel output (as of 1978) was produced by so-called minimills that employ electric arc furnaces and use iron and steel scrap as raw materials. The minimill industry has remained competitive with foreign producers and has more than doubled its percentage of total U.S. raw steel output since 1965.

54. Walter, "Structural Adjustment," p. 490.

55. "Persistent Pay Gap," *Wall Street Journal,* April 18, 1984, p. 1, based on U.S. Department of Labor estimates.

56. Walter, "Structural Adjustment," p. 485.

57. Richard D. Raddock, "Cyclical and Secular Developments in the U.S. Steel Industry," *Federal Reserve Bulletin* (Washington, D.C.: Federal Reserve Board, February 1981), p. 124.

58. See Office of Technology Assessment, *Technology and Steel Industry Competitiveness* (Washington, D.C.: U.S. Government Printing Office, 1980), pp. 194–207, for a description of new steel-making technology. Direct reduction converts iron ore (pellets, sinter, etc.) into sponge iron at temperatures well below the melting point of iron. This process differs from the conventional blast furnace process in that a solid metalized product is produced rather than molten iron, and a wide variety of reductants may be used in place of metallurgical grade coke.

59. Crandall, *U.S. Steel Industry,* p. 29.

60. "Bonneville Rates to Utilities Set to Increase 22%," *Wall Street Journal,* October 3, 1983, p. 21.

61. Charles River Associates, *Electricity Consumption for Primary Aluminum Production in the Tennessee Valley Authority Region,* CRA Report 624, October 1981 (reprinted in U.S. Congress, House, Committee on Public Works and Transportation, Subcommittee on Water Resources, *Hearings,* 97th Cong., 2d sess., April 30, 1982, pp. 154–64). A 1981 study prepared by Charles River Associates for the TVA concluded that TVA power rates would preclude any increase in power consumption at full capacity by primary aluminum smelters located in the TVA region.

62. Ibid., pp. 161–63.

63. For example, current Australian power rates are low because they are based on the cost of producing power from low-cost coal with few alternative uses. However, this situation is expected to change with the rise in demand for power in Australia.

64. "Power Costs Blamed for Deferring Plant," *Eugene Register Guard,* January 25, 1983, p. 6.

65. "1982 Survey of Mine and Plant Expansion," *Mining Activity Digest* 9, no. 8 (January 1982): 3.

66. "1983 Survey of Minc and Plant Expansion," *Mining Activity Digest* 10, no. 8 (January 1983): 45.

67. "1982 Survey of Mine and Plant Expansion," 4.

Chapter 6

1. The other eight core commodities are cocoa, coffee, sugar, cotton, jute, rubber, sisal, and tea; measures other than buffer stock agreements are proposed for bananas, beef and veal, rice, wheat, and wool.

2. The United States became a party to the Fifth International Coffee Agreement in 1964. It joined the International Tin Agreement in 1976.

3. Remarks by the Honorable C. Fred Bergsten, Assistant Secretary of the Treasury for International Affairs, before the Tenth Washington Conference for Corporate Executives of the Council of the Americas, *News Release* (Washington, D.C.: Department of Treasury, June 27, 1977). The U.S. position as outlined by Bergsten was that a common fund would be useful in the financing of buffer stocks after individual commodity agreements had been negotiated. However, the Carter administration would not support the proposal for a $6 billion fund under UNCTAD control that would: (*a*) be the principal source of financing individual commodity agreements; (*b*) finance activities other than the operation of buffer stocks; or (*c*) be used to intervene directly in markets not covered by an agreement.

4. For a discussion of the common fund controversy, see Bernard Blankenheimer, "The North-South Dialog and Its Bearing on U.S. Commodity Policy," *Special Study of Economic Change* (Prepared for use of the Joint Economic Committee, U.S. Congress, 96th Cong., 1st sess., December 17, 1980, vol. 9, pp. 392–98).

5. For a presentation of opposing views by economists specializing in commodity markets, see *Stabilizing World Commodity Markets,* ed. F. Gerard Adams and Sonia A. Klein (Lexington, Mass.: Lexington Books, 1978). The papers in this volume were presented at the Conference on Stabilizing World Commodity Markets sponsored by the Ford Foundation and held at Airlie, Virginia, March 1977.

6. See, for example, Charles River Associates, *The Feasability of Copper Price Stabilization Using a Buffer Stock and Supply Restrictions from 1953–1976* (Cambridge, Mass.: Charles River Associates, November 1977), pp. 1–6.

7. Alfred Maizels, *Selected Issues in the Negotiation of International Commodity Agreements: An Economic Analysis* (Geneva: UNCTAD, 1982).

8. "Tin Council Decides to Maintain Price for Metal Despite Enormous Supplies," *Wall Street Journal,* December 24, 1984, p. 16.

9. United Nations, General Assembly, *Resolution 3202* (S-VI), May 1, 1974.

10. Raymond F. Mikesell, *New Patterns of World Mineral Development* (London: British–North American Committee, 1979), chap. 1.

11. The World Bank Group consists of the International Bank for Reconstruction and Development (the World Bank), International Development Association (IDA), and the International Finance Corporation (IFC). The World Bank and IDA have the same staff and make the same kinds of loans except that IDA loans bear no interest and have fifty-year maturities. The IFC takes equity positions in privately owned enterprises in addition to making loans for projects in which there is some domestic or foreign private investment.

12. World Bank, *Annual Report 1978* (Washington, D.C.: World Bank, 1979), p. 21; see also National Advisory Council on International Monetary and Financial Policies, *International Finance, Annual Report 1977* (Washington, D.C.: U.S. Government Printing Office, 1977), p. 62.

13. This estimate was derived from World Bank *Annual Report*s for 1978, 1979, 1980, 1981, 1982, and 1983.

14. Data for these estimates were derived from International Finance Corporation *Annual Report*s for 1978, 1979, 1980, 1981, 1982, and 1983.

15. For a statement of the U.S. mining community's position, see Richard de J. Osborne, "International Banking Policies Urgently Need Reform," *American Mining Congress Journal,* June 6, 1984, pp. 10–11. A subsidized loan may be defined as one for which interest and other terms are more favorable to the borrower than those available in private international markets. World Bank loans bear interest equivalent to that on government-guaranteed obligations, but LDC borrowers cannot obtain loans in private financial markets at World Bank rates.

16. International Bank for Reconstruction and Development, *Articles of Agreement* (Washington, D.C.: World Bank, July 1984), p. 1. Article I of the Articles of Agreement of the International Bank for Reconstruction and Development states that a major purpose of the bank is to "promote private foreign investment by means of guarantees or participation in loans and other investments made by private investors; and when private capital is not available on reasonable terms, to supplement private investment by providing, on suitable conditions, finance for productive purposes out of its own capital, funds raised by it and its other resources."

17. Even U.S. mining companies operating in the Third World have accepted loans from international public lending agencies. For example, the Cuajone copper mine in Peru, owned by ASARCO, Phelps Dodge, Newmont, and Billiton (Dutch), was financed to the extent of $15 million by an IFC loan.

18. "U.S. Votes No at World Bank More Often under Reagan," *New York Times,* November 26, 1984. U.S. Executive Directors of the World Bank and the IADB have voted against loans to government mining enterprises in Chile and Zambia, among others. It is not known whether these actions were taken in response to pressure by private mining interests.

19. Glenn E. Blitgen, "U.S. Deepsea Mining at the Crossroads," *American Mining,* February 1982, pp. 51–54.

20. *The Future of Nickel and the Law of the Sea,* Mineral Policy Background Paper no. 10 (Ontario: Ministry of Natural Resources, 1980), p. 17.

21. Ibid.

22. Ibid.

23. National Materials Advisory Board, *Manganese Reserves and Resources of the World and Their Industrial Implications* (Washington, D.C.: National Academy Press, 1981), p. 48; see also J. P. Clark, N. J. Grant, and T. B. King, "The Market for Manganese Derived from Deepsea Nodules," *Natural Resources Forum,* July 1981, pp. 249–59.

24. John E. Tilton, *The Impact of Seabed Minerals: A Qualitative Analysis* (Laxenburg, Austria: International Institute for Applied Systems Analysis, 1983), p. 25.

25. Public Law 96-283, 96th Cong., 2d sess., June 28, 1980, Sec. 201.

26. Ibid., Sec. 118.

27. For the text of the LOS Convention, see United Nations, Third United Nations Conference on the Law of the Sea, *United Nations Convention on the Law of the Sea* (A/Conf. 62/122), October 7, 1982. The analysis of the convention in this section is based on a number of journal articles; see Bernardo Zuleta, "The Law of the Sea after Montego Bay," *San Diego Law Review* 20, no. 3 (April 1983): 475–87; David L. Larsen, "The Reagan Administration and the Law of the Sea," *Ocean Development and International Law Journal* 11, no. 3/4 (1982): 297–330, and other articles cited in this chapter.

28. United Nations, *Convention on LOS,* Annex III, Art. 13. During the first ten years the share to be paid to ISA is 35 percent on that portion representing a return on investment that is less than 10 percent; 42.5 percent of return between 10 and 20 percent; and 50 percent of return over 20 percent. In the second ten years of commercial production the corresponding shares to be paid to the ISA are 40 percent, 50 percent, and 70 percent, respectively.

29. "Recent Developments in the Law of the Sea, 1981–1982," *San Diego Law Review* 20, no. 3 (April 1983): 689. The parallel system of mining was first proposed by Secretary of State Henry Kissinger.

30. Ibid., pp. 694–95.

31. Elliot L. Richardson, "The United States Posture toward the Law of the Sea Convention: Awkward but Not Irreparable," *San Diego Law Review* 20, no. 3 (April 1983): 505–17.

32. "Recent Developments," pp. 701–4; for a discussion of the complex question of whether a country (such as France) that joined the convention could also operate under a reciprocating or minitreaty with the United States, not a party to the convention, see Tullio Treves, "The Adoption of the Law of the Sea Convention," *Marine Policy,* January 1982, pp. 11–13.

33. Ibid. However, it has been pointed out that a ruling by the ICJ could not enforce provisions of the treaty on nonsigning nations unless these provisions had become "customary international law."

34. Elliot L. Richardson, "The Politics of the Law of the Sea," *Ocean Development and International Law Journal* 11, no. 1/2 (1982): 21–22; see also Richardson, "United States Posture," pp. 505–17.

35. See statement of Elliot L. Richardson in U.S. Congress, House, Committee on Foreign Affairs, *Hearings, U.S. Foreign Policy and the Law of the Sea,* 97th Cong., 2d sess., June-September 1982, pp. 89–110.

36. Ibid., p. 91; see also, General Accounting Office, *Impediments to United States Involvement in Deep Ocean Mining Can Be Overcome* (Washington, D.C.: U.S. General Accounting Office, 1982).

37. Testimony of James L. Malone, U.S. Congress, House, Committee on Foreign Affairs, *Hearings, U.S. Foreign Policy and the Law of the Sea,* 97th Cong., 2d sess., June-September 1982, pp. 84–110.

38. "Statement of the American Mining Congress," in U.S. Congress, House, Committee on Foreign Affairs, *Hearings, U.S. Foreign Policy and the Law of the Sea,* 97th Cong., 2d sess., June-September 1982, pp. 216–42; see also testimony of Carl H. Savit, Western Geophysical Corporation on Behalf of the National Ocean Industries Association, U.S. Congress, House, Committee on Foreign Affairs, *Hearings, U.S. Foreign Policy and the Law of the Sea,* 97th Cong., 2d sess., June-September 1982, pp. 113–27.

39. "Victory at Sea," *Wall Street Journal,* January 11, 1983, p. 30.

40. Tom Alexander, "The Reaganites' Misadventure at Sea," *Fortune,* August 23, 1982, pp. 129–44.

41. Steven Chapman, "Underwater Plunder," *New Republic,* April 20, 1982, pp. 17–20; for another criticism of the LOS Treaty, see W. Scott Burke and Frank S. Brokaw, "Law at Sea," *Policy Review,* May 1982, pp. 71–82.

42. For a discussion of the loss of efficiency arising from provisions of the treaty, see Per Magnus Wijkman, "UNCLOS and the Redistribution of Ocean Wealth," *Journal of World Trade Law* 16, no. 1 (January/February 1982): 37–43.

43. This is the view of Ambassador Richardson. For a discussion of the issues involved, see Jonathan I. Charney, symposium in *Law and Contemporary Problems* 46, no. 2 (Spring 1983): 37–54; also see John S. Baily, Deputy High Commissioner of Australia to Canada, and James L. Malone, Assistant Secretary of State and Chairman, U.S. Delegation to the Third UN LOS Conference.

44. *Trade Policy in the 1980s,* ed. William R. Cline (Cambridge, Mass.: MIT Press, 1983), chap. 14. This issue has been discussed by C. Fred Bergsten and William R. Cline with respect to steel and a number of other sectors, such as agriculture and textiles, burdened by similar international trade restrictions and disputes.

45. I have deliberately excluded the LDCs as parties to the negotiation of a sectoral agreement; only the major developed countries concerned with a particular product or group of related products should develop the accord. Experience has shown that international agreements involving important national economic interests cannot be reached by a large number of diverse countries, but only by a few states with close political ties and a large stake in a successful outcome.

46. France, Sweden, and Switzerland have limited government stockpiles, while Japan has a government-sponsored and government-subsidized program for accumulating inventories by Japanese mineral and metal associations. France has a state-owned economic and strategic stockpile that was initiated in 1975 with a 1985 target of two-months' supply of normal imports for a number of minerals. Sweden's National Board of Economic Defense stocks a large number of minerals needed in the event of war or blockade. Neither Britain nor West Germany has a government stockpile of nonfuel minerals.

Chapter 7

1. This report was transmitted to Congress by the White House on April 5, 1982, under the title "The National Materials and Minerals Program Plan and Report to Congress, Pursuant to the National Materials and Minerals Policy, Research and Development Act of 1980"; see also Federal Emergency Management Agency, *Stockpile Report to the Congress, April–September 1979* (Washington, D.C.: U.S. Government Printing Office, 1979), pp. 1–4. For a history and critique of the U.S. stockpiling program, see Raymond F. Mikesell, *Stockpiling Strategic Materials: An Evaluation of the National Program* (Washington, D.C.: American Enterprise Institute, 1986).

2. The terms "strategic" and "critical" as applied to minerals have not been clearly defined by government agencies despite their universal employment. It has been suggested that "strategic" should relate to the probability of a supply disruption or price increase in a given nonfuel mineral and its expected duration, while "critical" should refer to the adverse impact on the civilian economy or defense industries that would occur if supply were interrupted or prices sharply increased. This definition was recommended by the General Accounting Office in a report entitled *Actions to Promote a Stable Supply of Strategic and Critical Minerals and Materials* (Washington, D.C.: U.S. General Accounting Office, June 3, 1982), p. 14.

3. U.S. Congress, Senate, Committee on Energy and Natural Resources, *Hearings, The President's National Materials and Minerals Program and Report to Congress,* 97th Cong., 2d sess., June 29, 1982, pp. 2–24.

4. The planning factors underlying the determination of stockpile goals for 1979 are described in Douglas P. Scott, "The Macroenvironment for 1979 Stockpile Goals" (Memorandum, Federal Emergency Management Agency, Washington, D.C.); FEMA is responsible for setting stockpile goals subject to review by various government agencies.

5. The Strategic and Critical Materials Stockpiling Act of 1979 requires that funds for acquisition of materials be authorized by Congress, including funds derived from disposal of excess inventories. Both disposals and new acquisitions must be reviewed by congressional committees.

6. The White House, Press Release, July 8, 1985; for an analysis of the President's proposal, see Raymond F. Mikesell, *Stockpiling Strategic Materials* (Washington, D.C.: American Enterprise Institute, 1986), pp. 27 and 68.

7. National Materials Advisory Board, *Considerations in the Choice and Form of Materials for the National Stockpile* (Washington, D.C.: National Academy Press, 1982). In 1982 a long-range program for examining the quality of stockpiled materials and their degree of deterioration was initiated.

8. Testimony of Simon D. Strauss, U.S. Congress, Senate, Committee on Energy and Natural Resources, Subcommittee on Energy and Mineral Resources, *Hearings, Strategic Materials and Minerals Policy,* 97th Cong., 1st sess., April 7, 1981, p. 32.

9. This recommendation was issued in mimeograph by the advisory committee, together with a background statement dated November 15, 1984.

10. A. Jordan and R. Kilmarx, *Strategic Minerals Dependence: The Stockpile Dilemma,* The Washington Papers (Beverly Hills, Calif.: Sage Publications, 1979), pp. 46–47. For example, in April, 1977, President Carter stated that his administration

would "work with Congress to assure that raw materials from our strategic stockpiles are available to meet supply disruptions during peacetime and to aid industry in evaluating future market conditions. These efforts should exclude the use of the strategic stockpile for purposes of general price stabilization"; General Accounting Office, *Stockpile Objectives of Strategic and Critical Materials Should Be Reconsidered because of Shortages* (Washington, D.C.: U.S. General Accounting Office, March 11, 1975), p. iv. This report to Congress stated that "The Congress may also want to study the advisability of broadening the strategic and critical materials stockpile concept to release material to meet short-term economic as well as national defense emergencies"; National Committee on Supplies and Shortages, *Government and the Nation's Resources* (Washington, D.C.: U.S. Government Printing Office, 1976), pp. 134–40. This report recommended serious consideration of a selective stockpile (e.g., chromium, cobalt, manganese, and platinum) to guard against supply disruptions in peacetime, but not to stabilize prices.

11. Congressional Research Service, *A Congressional Handbook on U.S. Materials Import Dependency Vulnerability* (U.S. Congress, House Committee on Banking, Finance and Urban Affairs, Subcommittee on Economic Stabilization, 97th Cong., 1st sess., September 1981); Office of Technology Assessment, *An Assessment of Alternative Economic Stockpile Policies* (Washington, D.C.: U.S. Government Printing Office, August 1976). This is both a technical analysis of economic stockpiling, including simulations, and a legislative and policy history of the concept.

12. The frequent strikes in the U.S. copper industry have had an important influence on world copper prices.

13. *The Resource War in 3D—Dependency, Diplomacy, Defense,* ed. James A. Miller, Daniel I. Fine, and R. D. McMichael (Pittsburgh, Pa.: World Affairs Council, 1980); J. A. Overton, "The Resource War: It Can't Be Won without Being Waged," *American Mining Congress Journal,* March 1981.

14. The price elasticity of demand for a commodity measures the sensitivity of demand to changes in its price. Price elasticity of demand also measures the sensitivity of price to changes in the amount available for consumption. Elasticity of supply measures the responsiveness of supply to changes in price. A low (high) elasticity of demand indicates that a small decrease in the amount available for purchase will result in a relatively large (small) rise in price. A low (high) elasticity of supply means that a large increase in price will result in a relatively small (large) increase in the amount supplied.

15. Congressional Budget Office, *Strategic and Critical Nonfuel Minerals: Problems and Policy Alternatives* (Washington, D.C.: U.S. Government Printing Office, August 1983), pp. x–xi. This report lists four mineral categories as substantially at import-disruption risk, namely chromium, cobalt, manganese, and the platinum-group metals. The report also analyzes the import supply situation for bauxite-alumina, copper, lead, and zinc, but concludes that "none of them appear to pose a major vulnerability risk."

16. These are antimony, columbium, ilmenite, industrial diamonds, rutile, tin, vanadium, and zinc.

17. Stockpile legislation permits releases in peacetime for use by defense industries, but not for civilian production.

18. The principal publications dealing with economic stockpiling are Michael

W. Klass, James C. Burrows, and Steven D. Beggs, *International Minerals: Cartels and Embargoes* (New York: Praeger, 1980); Office of Technology Assessment, *An Assessment of Alternative Economic Stockpiling Policies* (Washington, D.C.: U.S. Government Printing Office, 1976); National Commission on Supplies and Shortages, *Studies on Economic Stockpiling: Public and Private Stockpiling for Future Shortages* (Washington, D.C.: U.S. Government Printing Office, 1976), app. A, pp. 109–26.

19. There may be a number of contingencies for each material. These include the amount and duration of each disruption and the number of disruptions over a given time span.

20. In welfare economics this loss is known as the loss of "consumer surplus." Consumer surplus arises from the fact that consumers as a group pay a lower price per unit than they would have paid had the available supply been smaller. When imports are curtailed, some consumer surplus is lost.

21. The loss of consumer surplus would depend upon the nature of the imported material's demand curve.

22. Such adjustments can be made because social benefits tend to decline as additional amounts of materials are released from the stockpile.

23. This is a gross oversimplification of probability analysis, but it conveys the essential idea of determining a single probability coefficient for a series of contingency scenarios.

24. For an analysis of alternative approaches to determining the social rate of discount, see Raymond F. Mikesell, *The Rate of Discount for Evaluating Public Projects* (Washington, D.C.: American Enterprise Institute, 1977).

25. The concept of a "socially optimum" or "efficient" stockpile is analyzed in National Commission on Supplies, *Studies on Economic Stockpiling,* chaps. 1 and 4. "Socially optimum" or "efficient" is strictly an economic concept. It does not take into account income distribution effects.

26. The methodology is complicated in part because both social benefits from release of the last stockpiled unit and its social cost must be determined simultaneously, and both are a function of the optimum level of the stockpile itself, the determination of which is the objective of the calculation. Charles River Associates prepared a number of studies (most of them under contract with the government) that present both methodology and application in estimating optimal stockpile levels. Some of these studies are summarized in Klass, Burrows, and Beggs, *International Minerals;* for examples of mathematical stockpiling models, see National Commission on Supplies, *Studies on Economic Stockpiling,* app. A, pp. 109–26; see also Office of Technology Assessment, *Assessment of Alternative Policies,* chaps. 4 and 5; a highly simplified description of the methodology for determining the optimum stockpile level is given in Raymond F. Mikesell, "Economic Stockpiles for Dealing with Vulnerability to Disruption of Foreign Supplies of Minerals," *Materials and Society* 9, no. 1 (1985): 59–128.

27. Klass, Burrows, and Beggs, *International Minerals,* pp. 158–61. The estimates of optimum stockpile levels given by CRA and reported in this section for chromium, cobalt, manganese, and the platinum-group metals assume no *deterrent* effects from the existence of stockpiles or cartel actions. CRA gives somewhat higher stockpile estimates for the case in which deterrent effects are taken into account. I believe the possibility of cartel action with respect to these minerals is extremely low.

28. Ibid., pp. 202–3.

29. Average annual U.S. consumption for the 1979–81 period was 16.5 million pounds in contrast to about 19.5 million pounds assumed in the CRA study.

30. Klass, Burrows, and Beggs, *International Minerals,* pp. 178–79.

31. Ibid., p. 15.

32. For a discussion of the relationship between socially optimum stockpile levels and private-profit-maximizing inventories, see National Commission on Supplies, *Studies on Economic Stockpiling,* chaps. 3 and 4.

33. If releases from the stockpile were sold at the world price in the domestic market, exports and imports would continue to take place as dictated by price differentials between domestic and foreign markets, and stockpile releases would reduce the world market price. The contribution of stockpile releases to reducing U.S. payments transfers abroad applies to net transfers, so that any export of stockpile materials, or of products embodying them, would help to offset the cost of importing these materials.

34. Department of Energy, "Sale of Strategic Petroleum Reserve Petroleum," *Federal Register* 45, no. 246 (December 21, 1983): 56,538–42.

35. A. Silverman, J. Schmit, P. Oueneau, and W. Peters, "Strategic and Critical Mineral Position of the United States with Respect to Chromium, Nickel, Cobalt, Manganese and Platinum," mimeographed (Washington, D.C.: Office of Technology Assessment, June 15, 1983), chaps. 1 and 7. Goodnews Bay, Alaska, is the main domestic site where platinum is currently produced.

36. U.S. Congress, Senate, Committee on Energy and Natural Resources, Subcommittee on Energy and Mineral Resources, *Hearings, The President's National Materials and Minerals Program and Report to Congress,* 97th Cong., 2d sess., June 29, 1982; see also "AMC Supports DPA Funding," *American Mining Congress Journal,* April 27, 1983, p. 3.

37. U.S. Congress, Senate, Committee on Energy and Natural Resources, Subcommittee on Energy and Mineral Resources, *Hearings, The President's National Materials and Minerals Program and Report to Congress,* 97th Cong., 2d sess., June 29, 1982, pp. 163–77. Strong testimony in favor of the bill amending the Defense Production Act of 1950 was given by the president of Noranda Mining Inc., which holds a claim on the low-grade cobalt Blackbird ore body in the Salmon National Forest in Idaho.

38. H.R. 5540 was introduced in the 97th Cong., 2d sess., but did not pass. It was reintroduced as H.R. 13 in the 98th Cong., 1st sess., on January 3, 1983, but again its passage failed.

39. U.S. Congress, House, Committee on Banking, Finance and Urban Affairs, Subcommittee on Economic Stabilization, *Hearings, Defense Industrial Base Revitalization Act, Testimony on H.R. 5540 to Amend the Defense Production Act of 1950,* 97th Cong., 2d sess., March 23, 1982, p. 33.

40. Ibid.

41. Ibid., p. 57.

42. General Accounting Office, *Actions Needed to Promote a Stable Supply of Strategic and Critical Minerals and Materials,* Report to Congress (Washington, D.C.: U.S. Government Accounting Office, June 3, 1982), pp. 12–14.

43. Ibid., pp. 12–13. More on environmental benefit-cost considerations in chap. 8.

44. U.S. Department of Interior, *Cobalt: Effectiveness of Alternative U.S. Policies to Reduce the Cost of a Supply Disruption,* mimeograph (Washington, D.C.: Department of Interior, August 1981), chap. 3; in an unpublished internal memorandum, the Bureau of Mines obtained similar results in a simulation involving the subsidization of domestic chromium production.

45. U.S. Department of Commerce, *Critical Materials Requirements of the U.S. Aerospace Industry* (Washington, D.C.: U.S. Department of Commerce, October 1981), pp. 78–80.

46. Ibid., p. 81.

47. U.S. Department of Interior, "Aluminum," *Mineral Commodity Profiles* (Washington, D.C.: Department of Interior, May 1978), pp. 15–16; J. W. Smith, "Alumina from Oil Shale," *Mining Engineering,* June 1981, pp. 693–97.

48. K. Grjotheim and Barry Welch, "Impact of Alternative Resources for Aluminum Production on Energy Requirements," *Journal of Metals,* September 1981, pp. 26–32.

49. National Materials Advisory Board, *An Assessment of the Minerals and Materials Substitution Efforts of the Bureau of Mines* (Washington, D.C.: National Academy Press, 1983), pp. 15–16.

50. The National Strategic Materials and Minerals Advisory Committee appointed in April, 1984, is dominated by members associated in one way or another with the minerals industry.

51. The proposal would reduce the stockpile goal (calculated in 1979) from $16.3 billion for sixty-two materials (May, 1984, prices) to $6.7 billion. The NSC study involved scrutiny of the analysis, methods, and assumptions used by the previous administration. The review concluded that "A number of basic errors and unrealistic assumptions were used in the 1979 study."

52. Klass, Burrows, and Beggs, *International Minerals,* pp. 109–30.

53. The original members of the IBA were Jamaica, Suriname, Guinea, Guyana, Sierra Leone, Australia, and Yugoslavia. They were later joined by the Dominican Republic, Ghana, Haiti, and Indonesia. Brazil, which has large bauxite reserves and has been increasing its share of world output, is not a member of the IBA.

54. Alumina is fairly uniform in quality, but there are several grades of bauxite.

55. The Australian government does not impose an export tax on bauxite and has no desire to join a bauxite cartel.

56. The landed cost of bauxite required for producing one pound of aluminum represented only 14 percent of the 1981 U.S. producer price of aluminum. It would therefore require a substantial rise in export levies on bauxite to affect significantly the cost of producing aluminum.

57. Each alumina plant requires a certain mix of grades of bauxite for feed.

58. In 1981 the United States imported 890,000 tons of chromite ore and 315,000 tons of chromium ferroalloys. U.S. consumption of chromium from 1979 to 1981 averaged 543,000 tons per year. At the end of 1981, U.S. consumer stocks of chromite totaled 725,000 tons. In that same year recycled chromium amounted to 10 percent of total demand. Chromite ore contains 22 to 38 percent chromium, and chromium ferroalloys contain 36 to 70 percent chromium.

59. National Materials Advisory Board, *Contingency Plans for Chromium Utilization* (Washington, D.C.: National Academy Press, 1978), p. 2.

60. For a discussion of various types of vulnerability to supply disruption of chromium from South Africa, see Klass, Burrows, and Beggs, *International Minerals,* pp. 45–47; Leonard L. Fischman, *World Mineral Trends and U.S. Supply Problems* (Washington, D.C.: Johns Hopkins University Press for Resources for the Future, 1980), pp. 485–90; U.S. Congress, Senate, Committee on Foreign Relations, *U.S. Minerals Dependence on South Africa,* 97th Cong., 2d sess., October 1982.

61. National Materials Advisory Board, *Cobalt Conservation through Technological Alternatives* (Washington, D.C.: National Academy Press, 1983), p. 2.

62. For a discussion of cartelization possibilities, see Fischman, *World Mineral Trends,* pp. 478–80; and Klass, Burrows, and Beggs, *International Minerals,* p. 51.

63. National Materials Advisory Board, *Manganese Reserves and Resources of the World and Their Industrial Implications* (Washington, D.C.: National Academy Press, 1981), p. 71; Bureau of Mines, *Mineral Commodity Summaries 1980* (Washington, D.C.: U.S. Department of Interior, 1980), p. 94.

64. Fischman, *World Mineral Trends,* pp. 493–97.

65. Klass, Burrows, and Beggs, *International Minerals,* p. 222.

66. Ibid., p. 109. For example, if South Africa reduced sales, would the USSR respond with increased sales? And vice versa?

Chapter 8

1. See chaps. 5 and 7 on the issue of promoting domestic production to reduce vulnerability to import disruption.

2. This principle was first discussed by A. C. Pigou, *The Economics of Welfare* (London: Macmillan, 1920).

3. The Environmental Protection Agency requires eventual replacement of existing smelters if they cannot meet certain ambient air standards.

4. This is illustrated by the controversy over the 1985 closing of ASARCO's smelter in Tacoma, Washington, where copper concentrates containing arsenic and other dangerous materials were treated. Inadequate supplies of concentrates played a role in the decision to close the smelter—but so did the high cost of pollution abatement.

5. For a discussion of pollution's social costs, see Ronald Coase, "The Problem of Social Costs," *Economics of the Environment,* ed. Robert Dorfman and Nancy Dorfman (New York: Norton, 1977), 2d. ed., pp. 142–71; see also Ingo Walter, *International Economics of Pollution* (London: Macmillan, 1975), chap. 1; and M. H. Atkins and J. F. Lowe, *Pollution Control Costs in Industry: An Economic Study* (Oxford: Pergamon Press, 1977), chap. 1.

6. For an analysis of existing data on the relationship between sulfur dioxide levels and health, see Philip E. Graves and Ronald J. Krumm, *Health and Air Quality* (Washington, D.C.: American Enterprise Institute, 1981); see also Lester Lave and Eugene Seskin, *Air Pollution and Human Health* (Baltimore, Md.: Johns Hopkins University Press, 1977); and John Mullahy and Paul R. Portney, "Health and Air Quality—In Search of Missing Links," *Resources* 72 (February 1983): 6–7.

7. National Academy of Sciences, *Air Quality and Automobile Emission Controls* IV (Washington, D.C.: U.S. Government Printing Office, 1974). For example, a study published by the National Academy of Sciences assumed a value of $200,000

per death avoided in calculating the benefits of automotive air pollution control. One novel approach to determining the value of a worker's life is based on the trade-off between increased wages and increased probability of accidental death in particular employments. A recent article showed that wages will be 20 to 25 percent higher for each increase in the risk of accidental death at work (the increases are measured in terms of one death per thousand workers per year). Based on the increased wages paid for one thousand workers over the average working life, the study concluded that the monetary value of a worker's life was about $3 million. Allan Marin, "Your Money or Your Life?" *Three Banks Review* (London: Royal Bank of Scotland Group, June 1983), pp. 20–37.

8. Comptroller General, *Report to Congress: Cost-Benefit Analysis Can Be Useful in Assessing Environmental Regulations, Despite Limitations* (Washington, D.C.: U.S. Government Accounting Office, April 6, 1974), pp. 1–4. The USGAO argued strongly that the EPA should use cost-benefit analysis to improve data and methodology for such analysis.

9. Allen V. Kneese and Charles L. Schultz, *Pollution, Prices and Public Policy* (Washington, D.C.: Brookings Institution, 1975), pp. 87–96; see also Organization for Economic Cooperation and Development, *Polluter Pays Principle* (Paris: Organization for Economic Cooperation and Development, 1975), pp. 55–59; and Edwin S. Mills and Lawrence J. White, "Government Policies toward Automotive Emission Controls," *Approaches to Controlling Air Pollution,* ed. Ann F. Friedlaender (Cambridge, Mass.: MIT Press, 1979), chap. 8.

10. Raymond S. Hartman, Kirkor Bozdogan, and Ravindra M. Nadkarni, "The Economic Impact of Environmental Regulations on the U.S. Copper Industry," *Bell Journal of Economics,* Autumn 1979, pp. 589–618. There could be a problem with this approach, however, if ambient air quality standards applied to the region in which the smelters were located rather than to the actual emission of pollutants. By the use of high stacks a polluting source might protect the immediate region where it operates, but cause acid rain hundreds of miles away.

11. Robert W. Crandall, *Controlling Industrial Pollution: The Economics and Policies of Clean Air* (Washington, D.C.: Brookings Institution, 1983), pp. 54–67.

12. Meeting water quality standards costs a good deal for mining and milling and for steel and aluminum, but considerably less than meeting clean air standards.

13. A. Myrick Freeman III, "Air and Water Pollution Policy," in *U.S. Environmental Policy,* ed. P. R. Portney (Baltimore, Md.: Johns Hopkins University Press, 1978), chap. 2, pp. 31–34.

14. For a criticism of the technology standards approach, see ibid., pp. 45–49; and Crandall, *Controlling Pollution,* pp. 169–70.

15. N. D. Maniaci, "The Resource Conservation and Recovery Act: Potential Economic Impacts on Selected Mining Industries," *Journal of Environmental Science* 4, no. 6 (November/December 1981): 21–31.

16. Comptroller General, Report to Congress, *U.S. Mining and Mineral Processing Industry: An Analysis of Trends and Implications* (Washington, D.C.: U.S. General Accounting Office, October 31, 1979), p. 19.

17. For a statement of the American Mining Congress's position on exploration and mining in public lands, see "Declaration of Policy of the American Mining

Congress,'' *American Mining Congress Journal,* December 23, 1983. The Wilderness Act of 1964 provided for the location of mining claims and exploration in wilderness areas until the end of 1983. Deposits discovered by that date may be developed after 1983, but are subject to government regulation. Mining industry representatives have urged Congress to extend the 1983 cutoff date indefinitely.

18. James Baker, ''BLM Wilderness Review,'' *Sierra Club Bulletin,* March/April 1983, pp. 50–55. There are a number of mining claims within existing wilderness areas and National Park Service lands. The Sierra Club advocates federal acquisition of these claims for environmental protection.

19. Primitive areas constitute relatively small tracts that were so designated by the Forest Service, but have not yet been given wilderness status.

20. U.S. Congress, Senate, Committee on Energy and Natural Resources, Subcommittee on Energy and Mineral Resources, *Hearings, The President's National Materials and Minerals Program and Report to Congress,* 97th Cong., 2d sess., June 29, 1982, pp. 308–552.

21. For a discussion of various methods to evaluate benefits gained by visitors of wilderness areas, see Jack L. Knetscsh and Robert K. Davis, ''Comparison of Methods of Recreation Evaluation,'' in *Economics of the Environment,* ed. Robert Dorfman and Nancy Dorfman (New York: Norton, 1977), pp. 459–68; see also Marion Clawson, *Methods of Measuring the Demand for and Value of Outdoor Recreation* (Washington, D.C.: Resources for the Future, 1959).

22. For a discussion of this issue, see John V. Krutilla and Anthony C. Fisher, *The Economics of Natural Environments* (Baltimore, Md.: Johns Hopkins University Press, 1975), chap. 4; see also Anthony C. Fisher, *Resource and Environmental Economics* (Cambridge: Cambridge University Press, 1981), chap. 5.

23. Krutilla and Fisher, *Economics of Natural Environments,* chaps. 3 and 4.

24. Ibid., pp. 168–69.

25. AMAX's famous Climax mine, the largest molybdenum ore body in the world, has been shut down for several years because of a large surplus of the mineral. (Molybdenum is also a by-product of copper and other mineral mining and is in plentiful supply.)

26. Krutilla and Fisher, *Economics of Natural Environments,* pp. 168–69.

27. Ibid., pp. 184–86.

28. David Swan, ''Flexible Environmental Policy Needed for Primary Nonferrous Smelters,'' *American Mining Congress Journal,* January 1982, p. 26.

29. Arthur D. Little, *Economic Impact of Environmental Regulations on the United Copper Industry,* Report submitted to U.S. Environmental Protection Agency, Contract No. 68-01-2842 (Cambridge, Mass.: Arthur D. Little, January 1978). ADL concludes that because of the lead time required to install new control technology meeting EPA standards, U.S. smelter capacity expansion would be impossible by 1987. The analysis in the text of EPA regulations as they relate to the copper industry is based on the article by Hartman, Bozdogan, and Nadkarni, ''Economic Impact,'' pp. 589–618.

30. Bureau of Mines, *Mineral Commodity Summaries 1985* (Washington, D.C.: Department of Interior, 1985), p. 40; and William G. Siedenburg, *Copper Quarterly* (New York: Smith Barney, Harris Upham and Co., October 1985), p. 11.

31. Kennecott and Mitsubishi (Japan) have formed a joint venture for a $100 million expansion and modernization of the previously existing smelter, which was completed in early 1985.

32. Siedenburg, *Copper Quarterly,* January 18, 1983, p. 19.

33. Reverberatory and flash furnaces employ intense heat to produce a copper matte from concentrates. Flash smelters produce sulfur dioxide at concentrations that make it possible to produce sulfuric acid; it is not possible to capture SO_2 in reverberatory furnaces. Hydrometallurgical processes are chemical in nature.

34. Siedenburg, *Copper Quarterly,* July 1985, p. 21.

35. Frederick R. Demler, *Copper Market Outlook* (New York: Drexel, Burnham, Lambert, May 1985), p. 53.

36. For a discussion of various types of pollutants, see Organization for Economic Cooperation and Development, *Emission Control Costs in the Iron and Steel Industry* (Paris: Organization for Economic Cooperation and Development, 1977), chap. 4.

37. Comptroller General, Report to Congress, *Steel Industry Compliance Extension Act Brought About Some Modernization and Unexpected Benefits* (Washington, D.C.: General Accounting Office, September 5, 1984), p. ii.

38. House Committee on Energy and Commerce, *Steel Industry Compliance Extension Act of 1981* (Washington, D.C.: U.S. Government Printing Office, May 22, 1981), pp. 1–5.

39. Ibid., p. 9.

40. Ronald G. Ridker and William D. Watson, ''Long-Run Effects of Environmental Regulations,'' *Environmental Regulation and the U.S. Economy,* ed. Henry M. Pestin, Paul R. Portney, and Allen V. Kneese (Baltimore, Md.: Johns Hopkins University Press, 1981), p. 144.

41. OECD, *Emission Control Costs,* p. 95.

42. Robert W. Crandall, *The U.S. Steel Industry in Recurrent Crisis* (Washington, D.C.: Brookings Institution, 1981), pp. 39–40.

43. Crandall, *Controlling Pollution,* pp. 45–48.

44. Fluorides cause damage to livestock and vegetation in the immediate vicinity of primary aluminum smelter plants.

45. Environmental Protection Agency, *Primary Aluminum Draft Guidelines for Control of Fluoride Emissions from Existing Primary Aluminum Plants* (Washington, D.C.: U.S. Environmental Protection Agency, February 1979), p. 22.

46. Ibid., p. 10.

47. Ridker and Watson, ''Long-Run Effects,'' p. 144.

48. Department of Commerce, *The Effects of Pollution Abatement on International Trade—IV* (Washington, D.C.: U.S. Government Printing Office, November, 1976), p. 5.

Chapter 9

1. Donella H. Meadows, Dennis L. Meadows, Jorgen Randers, and William W. Behrens III, *The Limits to Growth* (New York: Universe Books, 1972). *Global*

2000 Report to the President, vols. 1, 2, and 3 (Washington, D.C.: U.S. Government Printing Office, 1980).

2. *The Resourceful Earth: A Response to Global 2000,* ed. Julian L. Smith and Herman Kahn (New York: Basil Blackwell, 1984).

Index